Jerzy Pogonowski

Essays on Mathematical Reasoning

Development in Humanities

edited by

Zbigniew Drozdowicz
(General Editor)

and Sławomir Sztajer

Volume 17

LIT

Jerzy Pogonowski

Essays on Mathematical Reasoning

Cognitive Aspects of Mathematical Research and Education

LIT

This publication has been financed by the Faculty of Psychology and Cognitive Science, Adam Mickiewicz University in Poznań

This book is printed on acid-free paper.

Bibliographic information published by the Deutsche Nationalbibliothek
The Deutsche Nationalbibliothek lists this publication in the Deutsche Nationalbibliografie; detailed bibliographic data are available in the Internet at http://dnb.dnb.de.

ISBN 978-3-643-91310-4 (pb)
ISBN 978-3-643-96310-9 (PDF)

A catalogue record for this book is available from the British Library.

Flössergasse 10
CH-8001 Zürich
Tel. +41 (0) 76-632 84 35
E-Mail: zuerich@lit-verlag.ch http://www.lit-verlag.ch
Distribution:
In the UK: Global Book Marketing, e-mail: mo@centralbooks.com
In North America: Independent Publishers Group, e-mail: orders@ipgbook.com
In Germany: LIT Verlag Fresnostr. 2, D-48159 Münster
Tel. +49 (0) 2 51-620 32 22, Fax +49 (0) 2 51-922 60 99, e-mail: vertrieb@lit-verlag.de

Contents

Preface

This volume contains four essays written in the last few years and related to the research project supported by the National Science Center research grant 2015/17/B/HS1/02232 *Extremal axioms: logical, mathematical and cognitive aspects*. Though not previously published, the essays are nevertheless connected with our recent major publication in the project, namely the book *Extremal axioms. Logical, mathematical and cognitive aspects*, Wydawnictwo Nauk Społecznych i Humanistycznych, Poznań 2019 as well as certain articles published in journals of local coverage. We believe that also presenting the ideas in question in LiT Verlag, a publishing house of greater international influence, could be expedient. Let us say a few words about the content of the essays:

Intuitive explanations. The paper contains a few reflections on the use of intuitive explanations in the context of transmission of mathematical knowledge, skills and abilities. Our remarks are based on an analysis of chosen mathematical textbooks as well as on our own teaching experience. Besides a tentative typology of such explanations, we are also interested in their effectiveness. We recall selected opinions concerning mathematical intuition expressed by mathematical educators.

A note on famous mistakes in mathematics. We discuss examples of famous mistakes in the history of mathematics. The term *mathematical mistake* is understood rather broadly here, counting as such not only lapses in calculations but, more importantly, wrong intuitions, hasty generalizations, false analogies, essentially incomplete proofs, and even certain conjectures that turned out to be false. The study of mistakes has to take into account their causes, the circumstances of their origin, the ways in which they have been corrected, their influence on the development of mathematics, and so on.

Standard, exception, pathology. We discuss the meanings of the concepts of *standard*, *exception* and *pathology* in mathematics. We take into account some pragmatic components of these meanings. Thus, standard and exceptional objects obtain their status from the point of view of an underlying theory and its applications. Pathological objects differ from exceptional ones – the latter may, for example, form a collection beyond some established classification while the former are in a sense unexpected or unwilling, according to some intuitive beliefs shared by mathematicians of the given epoch. Standard and pathology are – to a certain extent – flexible in the development of mathematical theories. Pathological objects may become "domesticated" and give rise to new mathematical domains.

Intended and standard models. The paper contains remarks on intended and standard models of mathematical theories. We recall some historical facts concerning such models. As examples we have chosen three mathematical structures: natural numbers $\mathbb{N}$, real numbers $\mathbb{R}$ and the universe of set theory. We mention the role of *extremal axioms* proposed in hope of a unique characterization of intended models. Finally, we also say a few words about mathematical intuitions connected with these models. In conclusion we stress the opinion that while intended models emerge as the result of mathematical research practice, standard models cannot be distinguished either on syntactic or semantic basis: in order to call a model standard we also have to take pragmatic criteria into account.

As should be visible from the above summaries our remarks belong to the philosophy of mathematics and are related mainly to mathematical *research practice*, in particular the ways in which we talk *about* mathematics. We are fully aware that the discussed topics are well-known to mathematicians themselves and as such they may be of lesser interest to them. However, we hope that the essays may attract the attention of those cognitive scientists who are interested in mathematical cognition.

Chapter 1

Intuitive explanations

1.1 Introductory remarks

The goal of this essay is very modest. We are not going to come up with any revolutionary theoretical results concerning mathematical intuition. Neither are we going to propose a coherent approach to teaching mathematics. But we nevertheless believe that the examples analyzed below and a couple of comments associated with them may have some relevance for reflections about the transmission of mathematical knowledge, skills and abilities. The term *context of transmission* was introduced in Pogonowski 2016 as an addition to two other well-known contexts, namely those of *discovery* and *justification*. In this essay we make essential use of material already presented in chapter 8 of Pogonowski 2019 as well as in the article Pogonowski 2018.

It should be also stressed that we use the term *explanation* in a popular sense and not in the sense attributed to it in the general methodology of sciences. The latter, theoretical approach to *explanation in mathematics* is discussed for example in: Steiner 1978, Kitcher 1984, Lange 2014, and Mancosu 2015. Our use of this term is related to questions of learning and teaching, when *to explain* means not only to recall the appropriate definition or refer to a suitable proof but also embraces several supplementary tools used in order to achieve a better understanding of the ideas discussed.

We actually think of our remarks as contributing to the broad subject of *argumentation in mathematics*. Argumentation involving intuitive ideas is of course ubiquitous in the teaching of mathematics. One can also find references to intuitive views in original research papers of professional mathematicians, although they are rather scarce. Here we limit ourselves mostly to the former.

We are not going to stubbornly defend the adequacy of the title of this paper. It is perceivable that a better title would refer to, say, *clarification* of mathematical ideas in teaching or to *heuristic methods* supplementing the understanding of mathematical concepts.

Obviously, our remarks below presuppose agreement with several results obtained so far. We appreciate, take for granted and make use of the achievements of Polya, Schoenfeld, Tall, Krygowska and many other researchers. In shaping our thoughts we especially owe a lot to the work Sierpińska 1994. We accept her approach to the understanding of mathematical notions. We also follow her in the belief that overcoming epistemological obstacles is one of the main problems in teaching mathematics.

Without going into detail we provide a brief recollection below of some views about the context of discovery and the context of justification in mathematics. The former is where one speaks of mathematical intuition as the source of invention. What are the connections between mathematical intuitions (of professional mathematicians) responsible for mathematical discoveries and the intuitive explanations used in textbooks in order to promote understanding of pre-established mathematical ideas? This question has at least two aspects. First, it concerns the ties between *invention* and *comprehension* of mathematical ideas. It seems that in order to grasp the meaning of a mathematical concept it is not necessary to follow exactly the way of thinking of its inventor. Moreover, due to the currently accepted style of writing, mathematical texts do not reveal the context of discovery of the results contained in them. Secondly, there may exist much easier (from a didactic point of view) ways of discussing the intended meaning of a particular concept than the way in which it was originally introduced by its inventor. A mathematical concept once brought into legitimate being in the realm of mathematics (that is, accepted by the mathematical community) interacts with other concepts, may be situated in new contexts, may generate a bunch of new facts, and so on. All this may broaden the possibilities of its intuitive explanations.

Quite obviously, the goal of intuitive explanations is to evoke a better understanding of the discussed mathematical ideas. But how can we be sure that such-and-such an explanation serves this goal appropriately? Plain translation of a symbolic formula into English should not be considered as a complete explanation. However, there should exist a strong preservation of meaning between the explained idea and its explanation. One might thus suppose that this procedure is partly based on analogy or metaphor. But notice that the idea which we want to explain intuitively is more complicated than the offered explanation: intuitive explanations are, as a rule, certain simplifications (in a sense, they are *converse metaphors*).

Our main claims concerning the subject under investigation may be summarized as follows:

1. We add *the context of transmission* (shortly: CT) to the two contexts already considered in literature, i.e. *context of discovery* (CD) and *context of justification* (CJ) in mathematics.

2. This context is responsible for the *transmission of mathematical knowledge, skills and abilities*. It is here where the learning and teaching processes take place.

3. Sections 2 and 3 are auxiliary. Beginning from section 4 of the paper we talk about *mathematical intuition* in CT.

4. An intuitive explanation, as viewed in CT, is a relative concept. We claim that *the* intuitive meaning of a mathematical concept is a chimera – it does not exist. Rather, we can talk about a *bundle* of intuitive explanations associated with a given concept. The proper meaning of a mathematical concept is given by its definition in a formal theory and by axioms of this theory. Such meanings are being *domesticated* (clarified, explained) by intuitive explanations in CT. The ultimate goal of such explanations is the evocation of *understanding*. The tools used for this purpose include: paraphrase, translation, analogy, metaphor, and model building.

5. In section 4 we recall in highly simplified form certain views already presented in literature. The only novelty lies in the fact that we are trying to confront those views with our modest teaching experience, that is with teaching mathematics to adult students of the humanities. This audience is far from easy to approach with mathematical ideas.

6. Section 5 is a survey of examples of a few types of intuitive explanation occurring in the context of transmission. We are not aiming at a complete exposition, and just want to provide some salient illustrations.

7. The last section contains reflections on possible verification of the effectiveness of intuitive explanations in CT. There is a huge amount of literature on this topic already, and our proposals are in no sense revolutionary. Again, it should be stressed that we are primarily interested in teaching adult students of the humanities.

1.2 Context of discovery

The context of discovery in mathematics includes, as a major part, the intuitions of professional mathematicians. One of the most beautiful descriptions of how things happen in the context of discovery in mathematics is the classic work Lakatos 1976. We assume that potential readers of this paper all know this work and that there is no need to recall its content here. Let us only observe that Lakatos promotes a quasi-empiricist approach to mathematics. Such views have recently been growing in popularity.

The book Byers 2007 suggests that one of the most important factors responsible for the development of mathematics is the everlasting struggle with antynomies and paradoxes. We may add that paradoxes do indeed play a regulatory role in the creative work of professional mathematicians: their resolution forces them to modify previously accepted intuitions.

The Encyclopedia of Philosophy (Edwards 1967) has the following entry concerning intuition:

> ***Intuition.*** The broadest definition of the term "intuition" is "immediate apprehension". "Apprehension" is used to cover such disparate states as sensation, knowledge, and mystical rapport. "Immediate" has as many senses as there are kinds of mediation: It may be used to signify the absence of inference, the absence of causes, the absence of the ability to define a term, the absence of justification, the absence of symbols, or the absence of thought. Given this range of uses, nothing can be said about intuition in general. Instead, it is necessary to pick out those principal meanings of the term which have played the most important roles in philosophical controversy and to discuss each of these individually. (Edwards 1967, volume 4, 204, article by Richard Rorty, 204–212)

The four principal meanings of the term "intuition" imply that this concept should be understood as: unjustified true belief not preceded by inference; immediate knowledge of the truth of a proposition, where "immediate" means "not preceded by inference"; immediate knowledge of a concept; and nonpropositional knowledge of an entity – knowledge that may be a necessary condition for, but is not identical with, intuitive knowledge of the truth of propositions about the entity.

The Stanford Encyclopedia of Philosophy enumerates several meanings of the term "intuition", including: belief which is not inferred from other beliefs; disposition to belief; intellectual conviction; and so on. Other encyclopedic sources contain similar characterizations.

We have discussed in brief selected views on mathematical intuition among philosophers (Descartes, Kant, Parsons, Tieszen) and mathematicians (Hilbert, Hadamard, Gödel, Thurston) in chapter 8 of Pogonowski 2019. For the purposes of this essay there is no need to repeat our comment here. In the next section, however, we recall selected views of educators which are related to our subject.

1.2.1 Properties of intuitive beliefs of professionals

In this section we will make some remarks about intuitive beliefs shared by the academic mathematical community. We think that it is unreasonable – at the present moment – to speculate about our cognitive abilities responsible for the mathematical intuitions of professionals. Instead we propose that mathematical intuition be understood as a complex of *verbalized* beliefs about the world of mathematics. We also prefer to talk about *intuition that* (something holds) rather than about *intuition of* (some object).

It is possible that not everything concerning the mathematical intuitions of professionals can be verbalized – after all, some activities of the mind may take place on the subconscious level. However, only explicitly stated judgments about intuitive beliefs can ever be put under investigation. Such judgments may come either from introspective reports of professional mathematicians or from an analysis of the source mathematical texts.

Mathematical intuition is presented in the axioms, which is evident: they are accepted without proof, solely on the basis of certain intuitive beliefs. One should keep in mind, however, that in the most important mathematical domains the axiomatic approach was preceded by a large accumulation of knowledge about the domain in question. There is an enormous amount of literature on this subject. Let us only mention, for example, that the paper Grzegorczyk 1962 discusses possible bases of justification for axioms of mathematical systems, while the paper Feferman, Friedman, Maddy and Steel 2000 tries to answer the question of whether mathematics needs new axioms.

We meet mathematical intuition also in the everyday practice of mathematicians – in reasoning by analogy, in making generalizations, in several heuristic rules of thumb, in using inductive assumptions before formulation of a hypothesis to be proven, and so on. Needless to say, the main body of mathematical education (especially teaching on the elementary level) is based on intuitive explanations.

Dynamics of intuitive beliefs

Unlike the more-or-less stable intuitions connected with everyday experience, mathematical intuition is more dynamic. In chapter 8 of Pogonowski

2019 we claimed that the major sources of changes in mathematical intuition seem to be:

Antinomies and paradoxes. Each time we find a contradiction in our mathematical considerations we are forced to discard it, which in turn implies rethinking the assumptions leading to this contradiction. Thus, for example, an allegedly innocent assumption accepted in the naive set theory that every property determines a set led to the Russell paradox. In order to save set theory from such contradictions, we had to change the formulation of the axiom of comprehension, which of course resulted in the construction of a new theory and in consequence implied also the corresponding change in our intuitions concerning sets. When resolving paradoxes, however, we do not change the theory but rather admit that the intuitions of everyday experience no longer apply at a certain level of sophistication of a mathematical theory. Paradoxes are thus more subjective in character than antinomies.

Scientific programs. Changes in intuitions can also be made deliberately, as a result of conducting some program. As examples one may list:

1. *Arithmetization of analysis.* The beginnings of differential and integral calculus were not based on sound logical ground (according not only to the current criteria of correctness, but also to those of that era). It was a great achievement of mathematicians of the 19th century (Cauchy, Weierstrass, Dedekind and others) to change this situation. Instead of thinking in terms of kinematic analogies (movement) and geometrical intuitions, it was proposed that analysis be based on arithmetical relations only. One should add, however, that infinitesimals were not expelled completely from mathematics at that time: Hahn, Stolz, Veronese and others studied several non-Archimedean structures in which infinitely small and infinitely large magnitudes occurred.

2. *Algebraic geometry.* Not very long after the first investigations in topology took place, topological properties began being described also using algebraic tools (for instance homology groups). In many cases it is much easier to characterize topological spaces via algebraic characteristics associated with them.

3. *Geometrization conjecture.* Thurston's geometrization program is an example of setting up new directions in mathematics. Out of the seven Millennium Problems proposed by the Clay Institute only one (the Poincaré conjecture) has been solved to date – and that quite recently. The problem itself is closely related to Thurston's program.

4. *Large cardinal axioms.* The axioms of set theory (Zermelo-Fraenkel first-order set theory) characterize the concept of set and the relation $\in$ rather weakly. Current work on set theory is directed partly by the search for new axioms which could change this situation. Typical examples of such new axioms are postulates requiring the existence of very large cardinal numbers (for instance strongly inaccessible, measurable, compact, and so on). Not all problems of set theory can be solved with such axioms (for example the continuum hypothesis could not be resolved in this way) but, still, large cardinal axioms appeared fruitful for instance in investigations of the consistency strength of theories. They also have many important applications in mainstream mathematics.

5. *Hilbert program.* The original Hilbert program required – roughly speaking – that we should prove the consistency of mathematics using finitary tools only, while also proving the completeness of mathematical knowledge. Due to the well-known incompleteness results (obtained by Gödel, Tarski, Turing and others) the program in this setting can only be partially realized. But the lesson learned was fruitful: we have changed our intuitive beliefs about intimate connections between proof and truth in mathematics.

6. *Hilbert's problems.* Hilbert's problems, which he laid out in 1900, influenced the mathematics of the entire 20th century. One only needs to look at the list of Fields medalists in the 20th century to see how much effort the mathematicians of this period put into resolving the problems from Hilbert's list.

New results. The accumulation of new knowledge on the domain in question is not without influence on our intuitions. It is often claimed that intuitive beliefs lack justification (and hence do not form knowledge, if we accept the view that knowledge is a justified true belief). This does not seem to be true in the case of mathematical intuitions. Probably any professional mathematician would agree that his intuitions become steadily deeper, more sublime and sophisticated as a result of his growing knowledge of the subject.

Mathematical intuition is influenced by (among others):

1. *Aesthetic values.* Professional mathematicians very often declare that their intuitions strongly depend on aesthetic values – theories, theorems, proofs, and constructions should be beautiful, in addition to the self-evident criteria of correctness. It is interesting that mathematicians share aesthetic judgments: this means that their opinions are not entirely subjective.

2. *Empirical experiments.* It is reasonable to assume that the origins of arithmetic lie in the processes of counting, the origins of geometry lie in measuring lengths, areas and volumes, and the origins of differential and integral calculus are influenced by reflections on motion, change, velocity, and so on. There always has been an interconnection between research in physics and mathematics. It is astonishing how many ideas of pure mathematics, developed allegedly without any connection to the physical world, have found – sooner or later – their applications in physics.

3. *Mathematical fashion.* This influence should not be underestimated. Each spectacular achievement in mathematics causes an increase of interest in the domain in question. Ambitious research programs also attract mathematicians to focus on specific subject in their work. The effects of these collective efforts are what direct the main streams of mathematics.

Sources of intuition

Two major sources of mathematical intuition seem to be:

1. *Our cognitive resources.* Our intuitions are most probably determined by the senses we have at our disposal. It is sometimes stressed that mathematical intuitions are mostly visual, partly connected also with touch and movement abilities. Quite what the basic mechanisms responsible for intuitive thinking are still remains a mystery. The results of several experiments do not explain much in this respect. One could also speculate how our mathematics would look, if we were – say – Rational Blots living in a completely liquid environment, with no access whatsoever to any rigid bodies. Should we then start with general topological spaces, with some kind of proximity idea instead of developing Euclidean geometry? One may construct several thought experiments of this kind, yet it seems to be but a speculative frolic. The real task is to explain genuinely existing mathematics and our mathematical intuitions as responsible for the growth of mathematical knowledge.

2. *Symbolic violence in the school.* Teaching mathematics is not based on memorizing. Rather, it is an interactive process of common problem solving, analyzing examples, or conducting constructions, and so on. At the beginning, a considerable role is played by the teacher's persuasion techniques. Pupils are guided to think in a way approved by the teacher, who is training them, correcting their mistakes and showing

the proper way to resolve the problems. It is often claimed that the main goal of teaching mathematics is just to evoke correct mathematical intuitions; the role of calculating abilities is only secondary.

As is well known, in a letter to Dedekind concerning the fact that the sets of points of $\mathbb{R}$ and $\mathbb{R}^2$ are equinumerous Cantor wrote the famous passage: *Je le vois, mais je ne le crois pas!* (I see it, but I do not believe it). Here *see* obviously refers to the fact that Cantor has obtained the *proof* while *believe* refers to his *intuitions* (and those of many of his contemporaries as well). After accepting the proof, Cantor was forced to change his intuitions (and so are we). In this case, it has been shown that cardinality is independent of dimension.

Sierpińska 1994 (page 36) recalls a confession made by Heisenberg in his *Der Teil und das Ganze* (1969), where he complained about his troubles with understanding of the theory of relativity: *Ich habe die Theorie mit dem Kopf, aber noch nicht mit dem Herzen verstanden* (I have understood the theory with my head but not yet with my heart). This case resembles the previously discussed situation of Cantor.

Let us finally (though marginally) add that *understanding* could be understood in different ways by different specialists:

1. Physicists claim that to understand (a physical phenomenon) means to build a mathematical model of it. In this case physical reality is to be explained, the basis of explanation is provided by mathematics, and understanding of mathematics is – we are guessing – self evident for physicists.

2. Researchers in the humanities claim that to understand a mathematical concept (idea, problem, etc.) means to present it in colloquial language. They also claim that to understand, say, a social phenomenon it is sufficient to explain it without any sophisticated mathematical language. Perhaps we are wrong in this suspicion, and not just with respect to many such researches. However, this work's author has on several occasions heard their declarations along the lines of: *it is enough that the students understand such-and-such a phenomenon when described in natural language, we can omit its mathematical description.*

1.3 Context of justification

Proof is a confirmation of intuition, as is often claimed in literature. Mathematical proof is, by definition, based on logical consequence. It does not, however, coincide with purely formal proof, that is proof as defined in formal logic. It is commonly claimed and believed that each mathematical proof can be, at least theoretically, transformed into a purely formal proof. From a practical point of view, however, such a transformation is hardly possible, if at all necessary. Nevertheless, mathematicians usually reach agreement when they are asked to decide whether a given argumentation can be considered as a correct mathematical proof.

1.3.1 Proof

The recently accepted style of publication of mathematical results follows Gauss rather than Euler. The obtained results are accompanied with more or less detailed proof, without any hints or comments concerning the path leading to the result.

Proofs can be evaluated, both quantitatively and qualitatively. One can investigate complexity of proofs from a purely formal point of view, as is done in proof theory, using tools from mathematical logic and recursion theory. But proofs can also be evaluated qualitatively, with respect to such properties as: deep, interesting, explanatory, and so on. It is not always clear how to propose an adequate measure in such cases, but this seems to be not only a question of taste and individual preferences.

The concept of explanatory proof is discussed for example in Mancosu 2015, where the views of Kitcher and Steiner are also taken into account. A proof may be considered explanatory if it refers to essential properties of the investigated objects. The explanatory character of a proof may be connected with such properties as for example unified perspective, simplicity or level of generalization.

The notion of mathematical proof has a stable core: its connection with logical consequence. The pragmatic aspects of mathematical proofs have recently been undergoing change. First, there are *collective* proofs, that is proofs obtained not by an individual but by a group of researchers, working over a longer period, as for instance in the case of the classification of finite simple groups. Secondly, some calculations required in a proof cannot be executed by an individual in any reasonable time, but can be nevertheless completed with some ease by a computer program.

The correctness of a mathematical proof depends on the agreement of the academic community. However, there are situations in which some mathematicians use a kind of a *private language* discussing a mathematical problem. One such situation, recalled by Alexander Bogomolny (on his famous Cut-the-Knot internet page) concerns a case Paul Erdős was discussing the solution of the Monty Hall problem with another mathematician so that the interlocutors reached agreement, though their conversation was completely incomprehensible to others (citing Cut-The-Knot):

> Andrew Vazsonyi begins his memoir with a story of his first encounter with the infamous Monty Hall problem. The solution he reluctantly arrived at went against his gut feeling. Curiously, his friend Paul Erdős had even more difficult time accepting the solution. He just grew upset on hearing Vazsonyi's explanations. A few days later, Erdős met R. Graham and they arrived at a solution that made everything perfectly clear to Erdős. He telephoned Vazsonyi to share the news and the solution. As the latter observed, "It made no sense at all. Apparently, Graham and Erdős had developed a private language that I wasn't privy to."

We will come back to the question of private language in mathematics later, discussing *negotiations about meaning* in the context of transition of mathematical knowledge.

1.3.2 Ways towards proof

A way to a proof may be thought of as a bridge between contexts of discovery and justification. Mathematical invention is not algorithmic, not even deterministic but one can identify decisive factors guiding a mathematician's mind from ideas to proofs. Let us just mention a few of them:

1. *Imagination*. In this respect, we would like to point to the role of counterfactual thinking, to asking question of the kind: *What if... ?*

2. *Insight*. In a sense, this is a synonym of intuition. We prefer the views of Davis and Hersh (Davis and Hersh 1981), according to which intuition emerges as a result of dealing with mental representations of mathematical objects.

3. *Generalization and abstraction*. It is a natural tendency in mathematical research to establish ever newer contexts, to look for more and more abstract objects, to look for formulations that enable us to see the investigated objects from a wider perspective.

4. *Analogy*. Thinking based on analogy (structural resemblance in a certain respect) enables us to apply methods effective in one domain to another one (or to create a new domain).

5. *Induction*. When one observes a regularity in a collection of facts, one usually aims at a hypothesis that this regularity holds without exceptions.

6. *Abduction*. Suppose a mathematician has a clear vision of a theorem they want to prove. Then they are looking for a collection of assumptions from which the theorem would follow.

1.4 Context of transmission

Our main subject is the transmission of mathematical knowledge, skills and abilities as presented in the textbooks and in personal oral teaching in the class. It is important to notice that the transmission itself is an *interactive* process, as seen either from the teacher's perspective or (more importantly) the student's. Learning mathematics is a cognitive process. It is goal-oriented: its ultimate goal is to achieve understanding of mathematical concepts, ideas, theorems, and procedures.

1.4.1 Opinions on learning and teaching mathematics

Judging from numerous reports from teachers of mathematics (in Polish elementary and secondary schools), the views based on Piaget's proposals are still considered actual. The works by Zofia Krygowska and her school (Krygowska 1967) are very popular in Poland. Among the didactic goals in mathematical education one usually mentions the following:

1. Developing calculation skills.

2. Developing symbol manipulation abilities.

3. Developing problem solving abilities.

4. Developing mathematical modelling abilities.

5. Developing (or shaping) mathematical intuitions.

The last task seems to be the most difficult. There are at least two reasons for this. First, it is difficult to describe precisely the very notion of mathematical intuition. Secondly, the initial creativity of children entering school is frequently subjected to the pressure of the rigid regulations of the mathematics curriculum, and as a result is partly lost in the process of education.

Let us recall a few opinions among educators about teaching mathematics. We restrict ourselves to brief citations only.

Piaget

Jean Piaget is a legendary figure in cognitive psychology. His ideas have influenced numerous researchers working in the domain of mathematical education. Piaget's most important idea concerns the stages of cognitive development:

1. *Sensorimotor stage* (0–2 years). Includes: simple reflexes, first habits and primary circular reactions, secondary circular reactions, coordination of secondary circular reactions, tertiary circular reactions, novelty, and curiosity, internalization of schemata.

2. *Preoperational stage* (2–7 years). Includes: symbolic functions and intuitive thought.

3. *Concrete operational stage* (7–11 years). Includes: understanding conservation and reversibility, improvement of classification skills.

4. *Formal operational stage* (11–16 years and onwards). Includes: development of abstract thought, logical thinking, skills oriented towards problem solving.

Vygotsky

Lev Semyonovich Vygotsky worked in developmental psychology. His theory takes into account the development of higher cognitive functions in children. He claimed that they emerge through practical activity in a social environment. Sierpińska summarizes some of his ideas in the following passage (Sierpińska 1994, 142–143):

> The genesis of concepts in a child, according to Vygotsky, is the genesis of his or her intellectual operations such as generalization, identification of features of objects, their comparison and differentiation, and synthesis of thoughts in the form of systems. The very same operations lie at the foundations of understanding. This explains why I have been attracted by Vygotski's theory of the development of concepts. The various genetic forms of these operations, discovered and described by Vygotsky, seemed to provide, almost immediately, the possible genetic forms of understanding. Moreover, the theory can be used to explain some of the curious ways in which students understand mathematical notions, and why, at certain stages of construction of these notions, they simply cannot understand in a different or more elaborate or more abstract way.

> Generalizations of mathematical objects and intuitions, identification of certain elements or features of these objects and situations as objects in themselves, or as more important than others, discrimination between objects and features of objects, synthesis of various judgments about objects and relations between objects, are the elementary operations both in understanding, and concept formation, and in thinking, in general. But, from a qualitative point of view, these operations are by no means the same in a child and in an adult. From early childhood through elementary school years through adolescence and adulthood, the operation of generalization, for example, undergoes an evolution from simple formation of chaotic aggregates of objects linked by various subjective and affective relations, to connections of objects on the basis of their playing a role in some common situation, to connection of objects linked by some common and abstract structure.

Polya

George Polya deeply influenced research in mathematical problem solving. The famous rules of problem solving from his work Polya 1945 are the following:

1. First, you have to understand the problem.

2. After understanding, then make a plan.

3. Carry out the plan.

4. Look back on your work. How could it be better?

Polya illustrates this simple program with numerous examples, in the work quoted, as well as in his later significant publications: Polya 2009, 2014. The above mentioned steps are further elaborated; in the case of plan making they include, for example (the list is not ordered):

Guess and check	Look for a pattern
Make an orderly list	Draw a picture
Eliminate possibilities	Solve a simpler problem
Use symmetry	Use a model
Consider special cases	Work backward
Use direct reasoning	Use a formula
Solve an equation	Be creative

Fischbein

Regarding the cognitive-behavioral function of intuition, Efraim Fischbein writes:

> In our opinion intuition is the analog of perception at the symbolic level. It has the same behavioral task as perception, namely to prepare and to guide our mental or practical activity. Therefore an intuitive conception must possess a number of features analogous to that of perception: globality, structurality, imperativeness, direct evidence, a high level of intrinsic credibility. In this way intuitions are able to inspire and guide our intellectual endeavors firmly and promptly even in a situation of uncertain or incomplete information. (Fischbein 1987, 201)

Schoenfeld

Alan Schoenfeld is one of the most influential figures in the world of mathematical educators. He is most famous for his works concerning mathematical problem-solving strategies. He insists that we should not limit our teaching to the transmission of mathematical content, but should also teach heuristics applicable in problem resolution.

According to Schoenfeld, the quality of our problem-solving attempts depends on our preexisting knowledge, which can be classified into the following categories:

1. our actual knowledge base,
2. our problem-solving strategies,
3. our control, monitoring and self-regulation (which we can call metacognition),
4. our beliefs and the practices that give rise to them.

Schoenfeld stresses also the following important advice: that students should always feel that the mathematical activities in which they are involved make sense.

Tall

David Tall has summarized a few decades of his experiences in mathematical education in the book Tall 2013. The most important idea of the book is that of *the three worlds of mathematics* suggested by the author. In actual fact, it tackles two independent classifications of mathematical activities. On the one hand we have the following division:

1. *Practical Mathematics*: with experiences in shape and space and arithmetic.

2. *Theoretical Mathematics*: with definitions based on known objects and operations.

3. *Formal Mathematics*: with formal objects based on formal definitions.

On the other hand, the author distinguishes:

1. *Conceptual embodiment.*

2. *Operational symbolism.*

3. *Axiomatic formalism.*

All these distinctions are illustrated by the diagram to be found at: http://homepages.warwick.ac.uk/staff/David.Tall/themes/three-worlds.html

The author evaluates his proposals in the following passage closing his book:

> The framework of three worlds of mathematics offers a broad picture of development of mathematical thinking from the newborn child to the frontiers of research through perception, operation and increasingly sophisticated forms of reason, taking account of supportive met-befores, that encourage generalization, and problematic met-befores, that impede progress. The theory formulates a framework for the long-term development of mathematical thinking that operates not only in the individual but also in the evolution of mathematics over the centuries. It also offers the possibility of blending with a range of other theoretical frameworks that encourage emerging insights that allow us to evolve more sophisticated insights.
>
> It reveals how the sensori-motor and linguistic capabilities of a biological brain evolve into the creative thinking processes of a mathematical mind. (Tall 2013, 419)

Sierpińska

We highly appreciate the work Sierpińska 1994, devoted to understanding in mathematics, mainly in mathematical education. The author characterizes the necessity of explanation in mathematics in the following words:

> The quest for an explanation in mathematics cannot be a quest for proof, but it may be an attempt to find a rationale of a choice of axioms, definitions, methods of construction of a theory. A rationale does not reduce to logical premisses. An explanation in mathematics can reach for historical, philosophical, pragmatic arguments. In explaining something in mathematics, we speak *about* mathematics: our discourse becomes more metamathematical than mathematical. (Sierpińska 1994, 76)

Sierpińska accepts Ajdukiewicz's theory of meaning and understanding. Ajdukiewicz wrote: *a person understands an expression if on hearing it he directs his thoughts to an object other than the word in question* (Ajdukiewicz 1974, 7). However, she extends it so that not only expressions may be understood, but also a problem. Thus *understanding* is, according to Sierpińska, a relation holding between *object of understanding* and *basis of understanding*. The *objects of understanding in mathematics* include: concepts, problems, formalism, and text. The *basis of understanding* includes: representations (e.g. mental images or conceptual representations, procedural representations), mental models, apperception, and thoughts that (so and so). Sierpińska stresses the fact that the mathematical objects are *seen as* something and not plainly *seen*. Understanding is considered as multidimensional, so to speak, because according to Sierpińska *mental operations* involved in understanding include: identification, discrimination, generalization, and synthesis. The necessary conditions for the act of understanding are *attention* and *intention*. The author points also to the role of *epistemological obstacles* in the process of understanding:

> In general, we propose to judge as more important than any other those acts of understanding that consist in overcoming an obstacle, whether developmental, or epistemological – related to the mature scientific knowledge.
>
> Let us take a simple example. When the notion of power zero is introduced by decreeing that $a^0 = 1$ some students will simply add this information to the mathematical rules they remember, without giving much thought to it. Others will accept it as a useful convention allowing to preserve the continuity of the exponential function $y = a^x$. But yet other students will revolt against it by saying that the very inscription a^0 does not make sense, if talking to power n means to multiply a by itself $n - 1$ time. In this case we say that conceiving of power as repeated multiplication is an obstacle to understanding the exponential function.
>
> It seems reasonable to admit that an understanding that is founded on a question or on an identification of something we do not quite understand, is superior to an understanding, unproblematic but banal

> and often merely verbal of $a^0 = 1$ as a convention and one more rule to remember. It goes deeper into the meaning of this convention, it starts a search for reasons. In fact, it implies a revision of the notion of power – an overcoming of an obstacle; it involves a construction and not just a memorization of the concept of real (a^x, x in $\mathbb{R}$). (Sierpińska 1994, 124)

Tao

Terence Tao distinguishes three stages in mathematical education (citing his internet blog Tao 2009):

> 1. The "pre-rigorous" stage, in which mathematics is taught in an informal, intuitive manner, based on examples, fuzzy notions, and hand-waving. (For instance, calculus is usually first introduced in terms of slopes, areas, rates of change, and so forth.) The emphasis is more on computation than on theory. This stage generally lasts until the early undergraduate years.
> 2. The "rigorous" stage, in which one is now taught that in order to do maths "properly", one needs to work and think in a much more precise and formal manner (e.g. re-doing calculus by using epsilons and deltas all over the place). The emphasis is now primarily on theory; and one is expected to be able to comfortably manipulate abstract mathematical objects without focusing too much on what such objects actually "mean". This stage usually occupies the later undergraduate and early graduate years.
> 3. The "post-rigorous" stage, in which one has grown comfortable with all the rigorous foundations of one's chosen field, and is now ready to revisit and refine one's pre-rigorous intuition on the subject, but this time with the intuition solidly buttressed by rigorous theory. (For instance, in this stage one would be able to quickly and accurately perform computations in vector calculus by using analogies with scalar calculus, or informal and semi-rigorous use of infinitesimals, big-O notation, and so forth, and be able to convert all such calculations into a rigorous argument whenever required.) The emphasis is now on applications, intuition, and the "big picture". This stage usually occupies the late graduate years and beyond.

He adds further comments claiming that: *The point of rigour is* not *to destroy all intuition; instead, it should be used to destroy* bad *intuition while clarifying and elevating* good *intuition.*

Our own position

As already stressed, our own teaching experience is restricted to mathematical education at university level. In addition, the audience of our classes comprises students in the humanities (first of all, in cognitive science). We would not dare claim that our reflections and proposals are applicable to a different, larger audience. Our position may be summarized as follows (citing Pogonowski 2019, 253–254):

> We propose adding a new context to the traditionally discussed contexts of discovery and justification, one which we tentatively name the *context of transmission* (of mathematical knowledge, skills and abilities).
>
> The context of discovery is responsible for the creation of new mathematical facts. The context of justification provides methods for establishing their validity. However, mathematics is also present in culture in different roles than just the creative work of professional mathematicians. There are at least five domains which should be taken into account here:
>
> 1. *Teaching mathematics.*
> 2. *Learning mathematics.*
> 3. *Popularizing mathematics.*
> 4. *Applications of mathematics in art and design, and so forth.*
> 5. *Philosophical reflection on mathematics.*
>
> All the above belongs to the context of transmission, which includes situations when we talk *about* mathematics. The first three of the mentioned domains are of special interest here. In all of them the focus is on the *transmission* of mathematical knowledge, skills and abilities. It is obvious that in order to achieve this goal we can apply tools which transcend mathematics itself. We find ourselves on the metalevel, so to speak. [...]
>
> In our opinion, the most important role in the context of transmission is played by *intuitive explanations*. We propose thinking about the problem as follows:
>
> 1. The meaning of mathematical concepts is determined by the underlying theory (ultimately, an axiomatic one).
> 2. Intuitive explanations connect that meaning with mental representations.
> 3. An understanding of a mathematical concept (such as a problem, proof, or construction) is obtained as a result of joining the

> (abstract) meaning together with a bunch of its intuitive explanations.

We distinguish several types of intuitive explanation, depending on their functional characteristic. Examples are provided in section 5.

1.4.2 The teacher's perspective

In the process of transmission of mathematical knowledge the teacher is supposed to influence the beliefs of others, namely the students.

The need to be persuasive is of utmost importance for any teacher. The teacher should be trusted. At the first levels of education she is an unquestionable authority. It is important, though, not to become authoritarian.

The didactic goals and recommended tools are constantly confronted with the teacher's own knowledge and vision of mathematics. They are of course dependent also on the stage of intellectual development of the audience.

We appreciate the idea (expressed by Sierpińska) that teaching mathematics includes *negotiations about meaning* of the discussed subjects. The roles of the participants of these negotiations are not symmetric but the important thing is that a didactic goal is achieved in an interactive process. The students are active in that process:

> First generalizations are built on images of the concrete and the finite. No wonder they resemble so much their archetypes. A point is a very tiny dot, infinity is 'so many that you cannot count' etc. Concepts cannot be given to the child, ready made, in the verbalized form or symbolic representation. The child has to construct them as generalizations of his or her previous generalizations and it is quite natural that the adolescent's first concepts may bear little resemblance to the fully fledged ones developed by generalizations made by mathematicians in their adult, mature, and often genius lives. And thus they become obstacles to understanding the theories. (Sierpińska 1994, 158)

If it happens that a teacher lives long enough then most likely they will sooner or later be faced with the drama of ever-changing recommendations for teaching styles. These changes are not solely connected with the development of mathematics itself. There are also other factors that influence such changes: the development of technology, the findings of cognitive psychology, ideological or political demands, and the pressure of public opinion.

Textbooks also accommodate these changes. The style based on the deductive method as in Euclid's *Elements* was replaced by a more intuitive-oriented style, then we experienced the rise and fall of the *New Math* program (influenced by the Bourbaki school), while more recently there has been vivid discussion (in the United States) about the *Common Core* program.

1.4.3 The student's perspective

The opinion that students are passive recipients in the process of transmission of mathematical knowledge is fundamentally wrong. Teaching is not like cattle breeding or carrot cultivation. We also strongly oppose the view that university is a place where students are clients and teachers are clerks offering knowledge packed in some kind of attractive container.

It is assumed that the fundamental goal among students is to fulfill their need for understanding. When a child enters elementary school he or she is full of curiosity about almost everything. Initially, the child is creative in a non-systematic way. The opinion that children lose *all* of their creativity due to school rigor is exaggerated. There are many factors responsible for success or failure in mathematical education. The stage of intellectual development, emotions, and social abilities all play crucial roles.

Due to the interactive aspect of mathematics teaching, the children get a sense of understanding of mathematical notions. One could say that they modify their private, idiosyncratic mental representations so that they finally reach (or only approach) representations unified with official standards.

We do not feel competent to discuss the subtleties of mathematical education in elementary or secondary school. Our target audience consists of adults whose school experiences are now behind them. Thus, their mathematical intuitions are to a certain degree regulated, they are not completely idiosyncratic. All the remarks that follow are related to this kind of audience.

1.5 Examples

In this section we provide some examples of intuitive explanations presented in mathematical textbooks. The sources are textbooks of advanced mathematics rather than elementary introductions.

1.5.1 Linguistic explanations

Natural language is the natural medium for teaching mathematics. Difficulties with understanding mathematical concepts are sometimes caused by the fact that students are required to associate a uniquely determined meaning with a term which also has a colloquial meaning, one well established in linguistic usage.

On the other hand, mathematicians prefer to use terms which somehow appeal to intuition. This can be visible for instance in topology, where one speaks of *dense* sets, *nowhere dense* sets, and so on. But the same can be said of any mathematical domain (analysis, algebra, arithmetic, etc.). Witty terms are appreciated in mathematics (for instance, the *overspill lemma* in arithmetic).

Baire's proposal of names for sets of the first and second category was rather unfortunate: part of the mathematical community prefers to talk about *meagre* sets (instead of sets of the first category).

While explaining mathematical notions some authors limit themselves to mere translations of symbolic notation into natural language expressions. Verbalization of a mathematical idea may be fruitful but it alone does not guarantee better understanding.

An example of a (rather funny) quarrel concerning linguistic explanations is the one related to Venn diagrams as representations of sets. One person insisted that there is a logically relevant difference between a situation where one set A in such a diagram is represented by a circle included inside a rectangle corresponding to the universe of discourse and a situation where this rectangle is simply divided in half by a line. This person said that the first representation means that A is *distinguished* (marked out) in the universe, while the second representation means that A is *in contradistinction* to the complement of A. Their disputant insisted, in turn, that the two diagrams were equivalent: they simply pointed to the distinction between A and its complement in the universe of discourse. In fact, the second of the Venn diagrams in question is exactly the same as the Carroll diagram representing this situation. Evidently the first person wanted to associate their diagrams with some extra meaning which they did not express by themselves, as his disputant rightly claimed.

Sets

Talking about sets in natural language is clearly necessary in the didactic process. However, one should be cautious, for example, with the distinction between distributive and collective meaning of the term *set*, with cases where a property allegedly determining a set is a fuzzy one, and with cases where a seemingly innocent property determines a proper class and not a set.

In everyday mathematical practice, as well as in elementary introductions to set theory, the universe of all sets is treated as precisely determined. The situation is of course different in advanced set theory, where several, sometimes bizarre, models of this theory are considered.

If the students have some pre-formal ideas (intuitions) about infinite sets, then most likely they would understand this notion as *potential* infinity:

a collection that can be ever expanded. The acceptance of *actual* infinity occurs not without resistance. The acceptance of different infinities becomes even harder, even when Cantor's diagonal argument is accepted.

It would be interesting to investigate whether a student's acceptance of the fact that the set of all real numbers has more elements than the set of all natural numbers is related to the order structure imposed on those sets. When we talk about sets, they are usually understood by the students as having a kind of internal structure: be it ordering, arithmetical dependencies, metric relations, and so forth.

The difference between dense and continuous ordering can of course be shown by referring to the formal definitions of these concepts. However, one may ask whether verbal presentations of these definitions suffice in order to understand their meanings. The difference in the cardinalities of $\mathbb{R}$ and $\mathbb{Q}$ is shown formally, but do the students, who operate with the metaphorical representation of these sets on the real line, understand this fact on the basis of proof or on the basis of this representation?

The powerset operation does not cause any problems in the case of finite sets. The first difficulty emerges when we talk for instance about the powerset of $\mathbb{N}$. From one point of view this is a well defined object. Yet, do we have *linguistic access* to it? Can all its elements be *named*? This is not a problem connected solely to natural language – in the first-order formal language of set theory we have at our disposal only a countable number of closed terms, and thus they are not sufficient to name all the elements of an uncountably infinite set.

The distinction between *defining* and *describing* is sharply visible when we talk about infinite ordinal numbers. There are several ordinal notations, but not all of the countable ordinal numbers can be described by them.

Students may be confused in the situation when the teacher says for instance: "Let T denote the set of all sentences true in the standard model of Peano arithmetic. We can prove that T is not definable in this theory." The teacher should explicitly refer to the difference between object language and its metalanguage, as well as explain what is meant by *not definable*.

Students are often confused by the fact that formulas from the object language and from the metalanguage appear in the same shape on the blackboard. Which of them are linguistic objects and which refer to the extra-linguistic realm?

Calculus

Let us now turn to the role of language in presentations of elementary analysis (Calculus) in textbooks. The ordered field $\mathbb{R}$ (with the norm based on absolute value) may be considered sufficient for developing Calculus. We make use of its algebraic and arithmetic properties (a number field), its order properties (continuous natural ordering compatible with field operations) and its topological properties (natural topology on $\mathbb{R}$).

The discussion about the structure of the continuum has a long tradition, going back to Aristotle. This is not the place to recall this story, but let us only mention that the views of mathematicians have changed several times, which can be seen for instance in the treatment of infinitesimal quantities in different epochs.

When teaching Calculus we must talk about $\mathbb{R}$ in natural language. What linguistic tools are appropriate for this purpose?

1. *Real variable.* It is customary to say that the independent variable somehow *moves* in the domain. Thus the student may think that it walks, runs, flies, swims (but doesn't jump) in this domain. All these metaphors are innocent as long as we remember that under the linguistic usage there is a precise meaning: the variable accepts (or takes) particular values in the domain.

2. *Limits.* It is customary to say that members of a sequence *tend* to its limit (if it exists). They *approach* the limit, are *arbitrarily close* to it. This is precisely expressed in the ε-δ setting, where closeness is characterized in terms of the absolute value.

3. *Continuity.* The fact that a function is continuous at a given point is linguistically expressed by saying that whenever a sequence of arguments tends to this point, then the sequence of values of these arguments tends to the value of the function at that point (the Heine definition). Again, this is precisely expressed in the ε-δ setting, where closeness is characterized in terms of the absolute value (the Cauchy definition).

4. *Derivative.* This concept is discussed with a reference to the *rate of change.* It is also clarified by the reference to geometry (tangent to a curve at a given point).

5. *Integral.* This concept is related to *summation.* It is customary to clarify it with a reference to the *area* below the curve.

In each of these cases it is important to stress explicitly that the linguistic tools used are imposed on us by the very fact that we communicate in a natural language. The transition between such figurative speech and understanding of mathematical concepts via their formal definitions is the mysterious core of the learning process.

Students may have trouble with all that ε-δ talk partly because it uses a string of quantifiers, which should be properly ordered (and cannot be omitted in the definition). But this formal complexity is the price paid for making Calculus free of extra-mathematical intuitive concepts.

One can encounter the opinion that teaching Calculus would be much easier in the framework of non-standard analysis. The definitions of the fundamental concepts become formally less complicated (a smaller number of quantifiers). However, the price for such simplicity should be paid in advance: the definition of hyperreals requires the construction of an ultraproduct, and it is debatable whether this task is an easy one, from a didactic point of view.

Geometry

Turning now to geometry, let us observe that Euclid's postulates look very innocent, so to speak, where their linguistic form is concerned. However, when it comes to *understanding* their meaning, a high level of abstraction is necessary:

1. A point does not have any parts.

2. A straight line (or rather, a segment – Euclid did not talk about infinite objects) has only length, without width. It "behaves evenly" with respect to all its points.

Such explanations demand that we transgress our perception of physical phenomena and enter the realm of purely abstract objects.

As is well known, there are systems of geometry in which one starts from more "friendly" objects as far as perception is concerned. Let us mention only two (there are numerous others): Huntington's system (with spheres and inclusion as primitive terms) or Tarski's geometry of solids.

We wish to end this subsection with two questions leaving the reflection of possible answers to the reader's amusement:

1. To what degree is natural language necessary in talking about mathematics? In other words, is it possible to teach mathematics at the purely formal level from the very beginning?

2. What are the limits of linguistic explanations in mathematics? In other words, do we have examples of mathematical concepts which escape a linguistic description?

1.5.2 Perception

Potentially, we could think of any kind of perception as relevant for intuitive explanations. However, the most pertinent types of perception (for the purposes under discussion) concern vision and touch (and perhaps also auditory and kinesthetic perceptions). Olfactory perception does not seem to be of any use here.

The introduction of numbers in early education is based on experiments with physical objects. This matter is widely discussed in literature, and we are not going to make any new contribution to the topic.

Pictures

Pictures, drawings, diagrams, and schemas are of course ubiquitous in the textbooks. In our opinion, they are totally legitimate in the context of transmission, provided that one is careful in following certain rules, for example:

1. Students should be notified that a picture is not a proof.
2. The pictures should not be too suggestive.
3. The conventions behind graphic representation should be consistent and explicitly given (say, in representing a three-dimensional object on a piece of paper).

As is well known, in the context of justification the status of such thing as pictures, drawings, and diagrams has been evo;vong through the history of mathematics. For Euclid (or Bhaskara) they were legitimate proof-tools. Some mathematicians rejected any suggestion of pictures being legitimate (notably Lagrange). "Thinking with diagrams" may be highly evaluated, for instance in algebra and category theory. As far as the context of discovery is concerned, it seems that the role of graphical (visual) representations depends on preferences and the style of thinking of individual professional mathematicians. We believe that no generalizations may be made in this respect.

The book Needham 1997 is an excellent example of how adequate pictures and diagrams can help in understanding even an advanced mathematical theory (in this case: complex analysis). Illustrations play a very important role in an introducing algebraic topology, see for instance Prasolov 2011. An interplay between diagrams and algebraic formulations is clearly visible in Adams 2004.

Poincaré wrote in *Science and Method* (citing Sierpińska 1994, 1):

> We are in a class of the fourth grade. The teacher is dictating: 'A circle is the position of the points in a plane which are at the same distance from an interior point called the centre.' The good pupil writes this phrase in his copy-book and the bad pupil draws faces, but neither of them understands. Then the teacher takes the chalk and draws a circle on the board. 'Ah', think the pupils, 'why didn't he say at once, a circle is a round, and we should have understood.'

This is an excellent example of what we mean by an intuitive explanation (in this case referring to visual perception). In an advanced course we may also stress the role of metric in the above definition, and show the students how a circle looks, say, in Manhattan metric.

3D models

I can still remember the rather rough 3D models used in the mathematical classroom half a century ago, when I was a pupil. Those were wooden, steel, plastic or plaster objects, representing solids discussed at school, such as cone, cylinder, ball, cube, and so on. Using such models at university level is probably considered not very serious, but I believe that is wrong. For example, one can illustrate the notion of the Lindenbaum-Tarski algebra with a model in, sat, cartoon format: it is the rhombic dodecahedron (in the case of algebra for formulas with two variables only; two vertices are "hidden" in this case). One can also imagine it as a tesseract.

I had the opportunity to attend a lecture by Istvan Lénárt, in which he showed us the many applications of a didactic tool set he had developed. The lecture was a resounding success – the audience seemed fascinated by his explanations (Lénárt 1996). The said set can be used , among others, to explain concepts in non-Euclidean geometry.

Movies, songs, color

Over the last few decades hundreds of videos discussing mathematical topics have been put on the internet. There are numerous lectures or lecture series devoted to particular domains or theories, recordings of panel discussions, and so on. But more importantly, one can speak of a new tendency in teaching mathematics – one based on (usually short) videos discussing a chosen topic. It would be fair to speculate that this form of teaching will continue to grow, partly because the first place where students of today look for information is the internet.

I used to end my lectures on mathematical problem solving with a math song found – not randomly – online. The didactic value of this method should not be exaggerated, but a good tune may further the chances of the discussed notions being remembered.

Colors are commonly used in elementary textbooks. However, they also appear useful in books on more advanced mathematics, notably in graphical representations of analytic functions (different colors are associated with different values of the modulus of a function). Color illustrations are expensive in printed text. But in texts produced electronically, as well as in computer animations, colors are "for free" and hence may be extensively used.

Tools

Experiments have demonstrated that our estimation of distance differs depending on whether it is vertical or horizontal. This seems to be related to our anatomical construction and to the gravitational effect.

We use several types of tools, instruments, and indicators in order to obtain data. The most widely known examples of mathematical instruments are (the choice is a little bit *ad hoc*):

1. *Compass*: used for drawing circles.
2. *Straight edge*: used for drawing straight lines or checking the straightness of lines.
3. *Ruler*: a straight edge with marked distances.
4. *Pantograph*: a mechanical linkage used for scaling a picture.
5. *Templates*: pieces of (solid) material with prescribed shapes.
6. *Protractor*: an instrument for measuring angles.
7. *French curve*: a template composed of many different curves.
8. *Lesbian rule*: a flexible tool made of lead that can be bent to a curve and used to measure or reproduce irregular curves.
9. *Slide rule*: a tool for calculating the results of arithmetic operations.
10. *Planimeter*: an instrument used to determine the area of an arbitrary two-dimensional shape.
11. *Trammel of Archimedes* (ellipsograph): a mechanical device used to draw ellipses.
12. *Tomahawk*: a tool used for angle trisection.

13. *Perspectograph* (plotter): an instrument used as an aid for drawing in perspective.

14. *Integraph*: an instrument used for plotting the integral of a graphically defined function.

15. *Spirograph*: a mechanical tool that produces mathematical roulette curves.

16. *Anamorphosis*: a distorted projection or perspective requiring the viewer to use special devices or occupy a specific vantage point to reconstitute the image.

17. *Opisometer*: an instrument for measuring the lengths of arbitrary curved lines.

18. *Calculators*: a term that now applies tp electronic devices used for many types of calculations.

19. *Abacus*: a calculating tool commonly used in different cultures.

In recent time some of these tools have become obsolete. They have been replaced by the possibilities offered by specialized computer programs. We did not include computers on the list above (either in their proto-forms or in their form existing today) – as this is a subject that goes far beyond the frame of this paper.

Software

Recent mathematical software such as *Matlab* or Wolfram's *Mathematica* offers astounding possibilities not only in research work itself but also in the learning and teaching of mathematics. Note that some textbooks even require the resolving of exercises via computer-based experiment (meaning: check something using the software). The free software *GeoGebra* also offers a lot of interesting tools for using in the teaching of arithmetic, algebra and geometry.

To a certain degree, the choice of mathematical notation may also be viewed as friendly (or unfriendly) for teaching purposes. One has to say, however, that attractive iconic notations do not always win in this race. For example Frege's notation for logical formulas, though consistent, was nevertheless rejected by the mathematical community. It was very complicated from a typographical point of view. On the other hand, Leśniewski's notation for sentential functors was typographically rather simple. Moreover, it was also very intuitive. Now it has almost been completely forgotten, though there are graphical methods of resolving semantic problems in the propositional calculus which resemble it, in a sense (the method of cross-diagrams).

1.5.3 Physics

The inspirations behind the creation of large portions of mathematics come from investigations in physics. It is thus natural to explain mathematical ideas by going back to their physical roots.

It would be difficult to provide a sharp distinction between *examples* from physics and *explanations* based on physics. Perhaps such an alleged distinction is an illusion, and when referring to physical situations in the teaching of mathematics we always work with examples.

The above problem is neither fictitious nor unimportant. It is related to the philosophical problem concerning the mathematical structure of physical reality. As the well-known *dictum* says, it is unexplainable why mathematics is so efficient in science. We are not obliged to reveal our philosophical standpoint on these matters but we would nevertheless like to express our belief that mathematical structure is somehow inherent in the physical reality. Of course, our personal views are of no significance whatsoever.

As it happens, people may have very bizarre images or representations concerning the behavior of physical bodies. For example, as Talia Ben-Zeev and Jon Star write (Ben-Zeev, Star 2001, 29):

> It has been found that when people were asked to draw the path of a moving object shot through a curved tube, they believed that the object would move along a curved (instead of a straight) path even in the absence of external forces.

Such views are classified as belonging to *folk physics*, also called *naive physics*. The views in question are very often simplifications or misunderstandings of real phenomena, and they are investigated in cognitive psychology. Whether of these views are innate is an interesting question.

When talking about intuitive explanations based on physics we would clearly like to exclude misconceptions of the above kind. That is to say we accept only such explanations that do not disagree with the current state of knowledge in physics.

Archimedes

Reference to physics can be found already in antiquity. When talking about the sphere, Euclid refers to *motion*: the surface of the sphere is obtained by rotating the semicircle around its diameter. But most notably such references can be found in the works of Archimedes. In the famous *The Method*, he describes a method to determine volumes which involves balances, centers of mass and infinitesimal slices. He is fully aware that such argumentation based on mechanics is not a genuine mathematical proof. He then provides strict proofs, using the method of exhaustion. To quote Archimedes himself:

> This procedure is, I am persuaded, no less useful even for the proof of the theorems themselves; for certain things first became clear to me by a mechanical method, although they had to be demonstrated by geometry afterwards because their investigation by the said method did not furnish an actual demonstration. But it is of course easier, when we have previously acquired, by the method, some knowledge of the questions, to supply the proof than it is to find it without any previous knowledge. This is a reason why, in the case of the theorems the proof of which Eudoxus was the first to discover, namely that the cone is a third part of the cylinder, and the pyramid of the prism, having the same base and equal height, we should give no small share of the credit to Democritus who was the first to make the assertion with regard to the said figure. I am myself in the position of having first made the discovery of the theorem to be published [by the method indicated], and I deem it necessary to expound the method partly because I have already spoken of it and I do not want to be thought to have uttered vain words, but equally I am persuaded that it will be of no little service to mathematics; for I apprehend that some, either of my contemporaries or of my successors, will, by means of the method when once established, be able to discover other theorems in addition, which have not yet occurred to me. (citing the text of *The Method*, Heath 2002, 13–14)

After this introductory exposition Archimedes writes down the assumptions concerning mechanical facts related to the center of gravity associated with geometric figures. There then follows a mechanical proof concerning the area of a section of a parabola. After this argumentation, Archimedes writes:

> Now the fact here stated is not actually demonstrated by the argument used; but that argument has given a sort of indication that the conclusion is true. Seeing then that the theorem is not demonstrated, but at the same time suspecting that the conclusion is true, we shall have recourse to the geometrical demonstration which I myself discovered and have already published.

The geometrical demonstration mentioned above uses the method of exhaustion. Historians of mathematics sometimes view Archimedes' usage of the method of indivisibles and the method of exhaustion as precursor work in differential and integral calculus. But opinions on this matter are not uniform. Anyway, this is not the subject of the present paper.

Mathematical Mechanic

Wonderful examples of intuitive explanations related to physics can be found in Levi 2009. Here is what the author writes about the intended audience of his book (Levi 2009, 4):

> If you are interested in mathematics and physics you will, I hope, not toss this book away.
>
> This book may interest anyone who thinks it is fascinating that
>
> - The Pythagorean theorem can be explained by the law of conservation of energy.
> - Flipping a switch in a simple circuit proves the inequality $\sqrt{ab} \leqslant \frac{1}{2}(a + b)$.
> - Some difficult calculus problems can be solved easily with no calculus.
> - Examining the motion of a bike wheel proves the Gauss-Bonnet formula (no prior exposure is assumed; all the background is provided).
> - Both the Cauchy integral formula and the Riemann mapping theorem (both explained in the appropriate section) become intuitively obvious by observing fluid motion.

In his comparison of the mathematical and physical approaches to solving problems, the author stresses their advantages:

1. In the mathematical approach we have rigor and universal applicability.
2. The advantages of the physical approach are:
   (a) Less or no computation.
   (b) Answer is often conceptual.
   (c) Can lead to new discoveries.
   (d) Less background is required.
   (e) Accessible to precalc students.

Physical argumentation is not supposed to *replace* mathematical proof: it should be considered only as an explanatory heuristic. Moreover, reference to physical models as intuitive explanations of mathematical ideas presuppose that the students know the corresponding physical phenomena (if not, then such explanation misses the point).

Problems concerning minima and maxima of functions seem to be well suited for physical explanations. After all, physical systems behave so that their states achieve extremal values. Just think of a pendulum finding minimal potential energy or a soap bubble finding the shape with least volume. The author gives the following schematic correspondence between calculus and mechanics which he later uses for mechanical proofs:

Calculus	**Physical interpretation**
The function $f(x)$	Potential energy $P(x)$
The derivative $f'(x)$	The force $F(x) = -P'(x)$
$f(x)$ minimal $\Rightarrow f'(x)$	$P(x)$ is minimal $\Rightarrow F(x) = 0$ (equilibrium)

The crucial thing in the examples of physical argumentations associated with a given mathematical result is choosing an adequate model. For example, in results from complex analysis (like, say, the Cauchy Integral Formula or Riemann Mapping Theorem) it could be the fluid flow or heat flow. Center of mass may be useful in argumentations associated with theorems in geometry (remember Archimedes?), effects connected with gas pressure can be helpful in illustrations related to curvature problems (Gauss-Bonnet Theorem), and so on.

Topology and linkages

Robert Ghrist has supplemented his *Elementary applied topology* (Ghrist 2014) with numerous interesting examples showing the interpretation of topological notions. However, we feel that the unprepared reader should be warned: the word *elementary* in the title refers to facts which are indeed elementary, but to the mathematician with some experience. Homology, cohomology, Morse theory, sheaves, and the like require some previous training in mathematics before they (and their applications) can be fully appreciated. We shall limit ourselves to quoting only one result from the book:

> **Theorem 1.4.** *Any smooth compact manifold is diffeomorphic to the configuration space of some planar linkage.*
>
> This remarkable result provides great consolation to students whose ability to conceptualize geometric dimension greater than three is lim-

> ited: one can sense all the complexities of manifolds *by hand* via kinematics. (Ghrist 2014, 12)

Linkages themselves have a great and interesting history. Archimedes, Hero of Alexandria, and Leonardo da Vinci were precursors of the investigations of the connection between geometry and mechanics. Linkages still play an essential role in engineering: in the construction of engines, robots, and so on. From a mathematical point of view, they are related to the problems of converting one kind of motion into another – for instance motion in a straight line into rotation or vice versa. This was a problem important for steam engines and sewing machines, among others.

Mathematicians developed several kinds of linkages, for instance:

1. *Peaucellier-Lipkin linkage.* Invented in 1864, was the first true planar straight line mechanism, meaning the first planar linkage capable of transforming rotary motion into perfect straight-line motion, and vice versa.

2. *Chebyshev's Lambda Mechanism.* This converts rotational motion into approximate straight-line motion with approximate constant velocity.

3. *Four-bar linkage.* The simplest movable closed chain linkage.

4. *Leg mechanisms.* Several types of linkage intended to simulate the walking motion of humans or animals.

We consider it a pity that the modern style of teaching mathematics at university does not – as far as we know – make use of these remarkable devices, at least for illustrative purposes. We can only speculate that the situation is different at technical universities (institutes of technology) conducting intense work in robotics.

Other examples

One could of course give endless examples of reference to physics in mathematical textbooks. However, we shall round off with just few more items.

Surprises. Nothing is more valuable in mathematical education than surprises, of which there are several types. Galperin 2003 analyzes the following experiment:

> Let the mass of two balls be M and m, respectively. Suppose that $M = 100^n m$ ($n \geqslant 0$). We roll the ball M towards ball m which is near the wall. Thus M hits m which bounces off the wall. How many collisions occur (jointly, i.e. between M

> and m and between m and the wall) before the ball M changes direction? The answer depends on n, of course.

A surprising fact is that the number of balls' collisions is equal *precisely* to the first $n + 1$ digits of π. It is worth noticing that the result is purely *deterministic* and not based on probability. Another interesting feature of this problem lies in the fact that changing a perspective (in this case: from plain description of motions to the configuration space) essentially simplifies the solution of the problem.

Harmonic series. The fact that the harmonic series is diverging can be used for the solution of the famous *ant on a rubber rope* problem (see for instance Havil 2007), which may be worded in brief as follows:

> An ant starts to crawl along a taut rubber rope 1 km long at a speed of 1 cm per second (relative to the rope it is crawling on), starting from its left fixed end. At the same time, the whole rope starts to stretch with the speed 1 km per second (both in front of and behind the ant, so that after 1 second it is 2 km long, after 2 seconds it is 3 km long, and so on). Will the ant ever reach the right end of the rope?

This example is an excellent opportunity for discussing the difference between a perfectly correct mathematical result and its realization in physical reality. On the other hand, the divergence of the harmonic series is used also in many very practical situations: displacement of resources on a route of a planned journey, testing of materials, choosing candidates applying for posts, and so on.

Paradoxes of motion. Scholten and Simoson 1996 contains a discussion of the *falling ladder paradox*, which can be concisely described as follows:

> A ladder of length L is leaning against a wall. The bottom of the ladder is being pulled away from the wall horizontally at a uniform rate v. Determine the velocity with which the top of the ladder crashes to the floor.

If one assumes that the ladder does not lose contact with the wall while sliding, then the alleged solution to this problem implies a physical impossibility: the ladder hits the floor with infinite speed. Actually, the ladder does lose contact with the wall at some moment, and its motion after that is described by the equation for the motion of a pendulum. We consider such examples as instructive, because they show that in some situations we are forced to replace our first mathematical approximation of the problem to a more accurate one. Further examples can be found for instance in Klymchuk and Staples 2013.

Variational problems. One of the first problems in the calculus of variations is that of *brachistochrone*, solved by Jean Bernoulli (and formerly inaccurately solved by Galileo). A nice intuitive exposition of Bernoulli's solution can be found in Polya 2014 (152–155). Intuitive explanations of some other variational problems may include for instance *minimal surfaces* – the mathematical description can be accompanied by physical models (say, soap bubbles).

Probability experiments. People often think of probability as an absolute concept. It is instructive to make clear to the students that probability depends on a pre-accepted measure. The Bertrand paradox about the probability of a randomly chosen chord of a unit circle being longer than the side of an equilateral triangle inscribed in the circle (which surprisingly to newcomers may be equal $\frac{1}{2}$, $\frac{1}{3}$ or $\frac{1}{4}$, depending on the choice of the space of elementary events) is a useful example in this respect. Recalling Buffon's experiment about tossing a needle (and determining the probability of it crossing parallel lines drawn on the paper) may make students reflect on how the famous mathematical constant π is involved in an empirical process.

Infinity and physics. Supertasks have been causing trouble from the very beginning of scientific (or even philosophical) reflection about Nature – cf. Zeno's paradoxes of motion. They are still vividly discussed, including in new forms (for instance: Thomson's lamp, Turing machines making an infinite number of steps in a finite amount of time, and so on). An interesting case of Laraudogoitia's colliding balls (Laraudogoitia 1996) is analyzed in Romero 2014 where its physical non-realizability is shown with a reference to the general relativity theory. We consider the discussion of supertasks as a very valuable tool for reflections on connections between mathematics and reality.

1.5.4 Common sense

Examples from this subsection are thought of as related to human behavior, to common-day phenomena, to experiences to which we are accustomed. They could also be classified among the previously discussed examples connected with perception.

References to everyday experience should be easily understood but they should be used carefully. Due to, among other things, several cognitive biases, common sense reasoning may diverge dramatically from mathematical inferences.

It is customary in elementary topology to refer to the investigated spaces as *rubber-like objects* and to mappings (homeomorphisms) as continuous deformations of such objects. Models made of, say, plaster can be used to illustrate such deformations (see for instance Prasolov 2011).

Students are of course familiar with references to *rigid motions* in elementary geometry. Translations and rotations can be easily illustrated in the case of a plane, but for instance rotations in space are a little bit more complicated.

Geometrical representations of natural numbers are used for example in Conway and Guy 1998, where several summation formulas are illustrated by suitably chosen diagrams depicting triangular, quadratic, pentagonal, hexagonal numbers, and so on. Formulas such as:

1. $1 + 2 + 3 + \ldots + n = \frac{n \cdot (n+1)}{2}$
2. $1 + 3 + 5 + \ldots + (2n - 1) = n^2$

are usually proven by mathematical induction. Geometrical illustrations accompanying such proofs may contribute to a better understanding of the investigated processes of summation.

Since recently, negative numbers have been discussed quite early on mathematical education. Several types of intuitive explanations are used in this context. One may refer, for instance to: debts, floors, temperature, game scoring, or jumping frogs, and so on. Even the fact that negative times negative gives positive, which for some pupils is difficult to understand, can be persuasively illustrated, as it can be seen in instructive videos accessible in the internet.

Contemporary students are better acquainted with computers and software than with books. Thus it seems reasonable to refer – if possible – to the action of computer programs when explaining mathematical concepts and procedures. For example, some computer games use the computer screen as if it were the surface of a torus and so one can use this fact when talking about topology.

Several mathematical constructions, notions, and procedures can be interpreted in terms of games. Of course, there exists a separate domain of mathematics dealing with games, winning strategies, and so on. Certain parts of mathematics (notably descriptive set theory) owe a lot to game-theoretical interpretation. Interesting examples can also be found in logic and model theory, for example: game-theoretic semantics, representations of proofs, Ehrenfeucht's games used in the characterization of the completeness property, and so on.

Intuitive explanations in algebra or topology can refer to our perception of size. For example, *filters* are often said to consist of *big* elements, while *ideals* are consisted of *small* elements.

Whether some more sophisticated notions could be given simple intuitive explanations is debatable. For instance, concepts from measure theory may be such. *Almost everywhere* (*almost always*, and the like) is difficult to be given a precise meaning in finite domains. *Almost always* in such a case should mean *always, with a negligible number of exceptions*. Then a numerical characterization of *ignorable* should be specified, which can be troublesome. *Almost all* in a countable domain (like $\mathbb{N}$) means *all with the exception of a finite number of elements of the domain*. Similarly, *almost all* in an uncountable domain (like $\mathbb{R}$) means *all with the exception of a countable number of elements of the domain*. This is quite different from the finite case. Students should be notified of this difference.

An example of figurative style, referring to human abilities, can be found in numerous works of set theory, where the authors use the phrase: "a model thinks". This expression means that some sentence is true in the model or that the model itself has some property, expressible in the underlying metatheory. For example, if one writes that model M thinks set A is uncountable, then this means that inside model M there does not exist a bijection between A and the set of natural numbers in M. Looking "from outside" M, set A may well be countable.

Thought experiments in a sense bear a stigma of intuitive explanations. They appeal to human imagination. They may conform to physical reality or even transcend it. However, it seems that we always require a certain degree of rationality when talking seriously about thought experiments: we exclude fairytale talk from this domain. The question remains as to how much rationality is included in common sense views.

1.5.5 Cross-domain explanations

Despite the fact that modern mathematics has more than three thousand subdisciplines, we still feel (or are even convinced) that it is coherent. Moreover, the mutual connections between different domains of mathematics happen to be very productive. Applying algebraic tools in topology enabled us to solve difficult topological problems via transformation into (as a rule, much simpler) algebraic problems. Some facts from the theory of functions of a real variable can be better understood when viewed from the perspective of complex analysis. These facts may suggest that we could successfully explain ideas from one mathematical domain by reference to another domain, presumably better understood by the students themselves. Such explanations are, in our opinion at least, most interesting. We have not yet collected many of them, and below we give only a few examples (discussed already in Pogonowski 2018 and are partly reprinted below).

Coherence of mathematics

Algebraic concepts have recently come to penetrate all domains of mathematics. This is an example of success achieved rather quickly – modern algebra has its origins only in the 19th century. But this algebraic trend is not the only testimony to the coherence of mathematics. In the day of Greek mathematics, geometrical investigations played the dominant role. There were of course arithmetical considerations as well, although the theory of proportions in Euclid's *Elements* could be thought of as subordinate to geometry. Descartes, with his origin of analytic geometry, gave a strong impulse for tightening geometry with algebra. Arithmetization of analysis in the 19th century was another step towards unification. More recently, one may think for instance of algebraic logic as a tendency to unify logical investigations with mainstream mathematics.

All such unifications also imply mutual influences between mathematical domains. Generalizations of arithmetic operations and relations to new domains (for example matrices or function spaces) may be thought of as examples of such cross-domain influences in mathematics.

Riemann hypothesis

Davis and Hersh discuss an interesting case concerning cross-domain explanation related to the Riemann hypothesis (Davis and Hersh 1981). Let us recall that the Möbius function $\mu(x)$ is defined as follows for any natural number x:

1. $\mu(x) = 0$, if x is divisible by a square of some number;
2. $\mu(x) = 1$, if all the factors in the prime factorization of x are different and there is an odd number of them;
3. $\mu(x) = -1$, if all the factors in the prime factorization of x are different and there is an even number of them.

Now, let $M(x)$ equals the sum of all $\mu(y)$, for $y \leqslant x$. It is known that the Riemann hypothesis is equivalent to some condition describing the rate of growth of $M(x)$ as compared with $x^{\frac{1}{2}+\varepsilon}$, where $x \to \infty$ and $\varepsilon > 0$. There comes an *intuitive* assumption: let us think of $M(x)$ as a random variable function. Indeed, it seems that prime factorizations may contain even or odd number factors, without any regularity, and hence the function μ takes the values 1 and -1 with equal probability. Omitting the calculations that follow, let us proceed directly to the conclusion. It claims that the Riemann hypothesis is true *with probability one*. For some readers this may sound reasonable (under the accepted probabilistic interpretation of the function M), but others may consider such a statement totally absurd.

Generic sets and forcing

We find also an interesting cross-domain explanation in Peter Koepke's lecture notes *Models of set theory I* (Koepke 2013). It relates the concepts of forcing and generic models to certain concepts from algebra concerning field extensions. First, let us recall that:

1. The field of complex numbers is algebraically closed. It contains the prime field of rational numbers $\mathbb{Q}$.
2. The field axioms do not decide the existence of, say, $\sqrt{2}$. It does not exist in $\mathbb{Q}$, but exists in $\mathbb{Q}(\sqrt{2})$. The number $\sqrt[3]{2}$ does not exist in field extensions of $\mathbb{Q}$ by square roots.
3. Given a field k, its extensions can be described within it:
   (a) If a is algebraic over k, then one can treat a as a variable x which is zero of some polynomial in $k[x]$.
   (b) If a is transcendental over k, then a corresponds to a variable x such that $p(x) \neq 0$ for all $p \in k[x]$.
4. Taking into account the countable field $\mathbb{Q}$, one can construct a transcendental real number $a = 0, a_0 a_1 a_2 \ldots$ by successively choosing decimals a_i in such a way that $0, a_0 a_1 a_2 \ldots a_m$ forces $p_n(a) \neq 0$, where (p_n) is some enumeration of the polynomials from $k[x]$. In other words, this choice implies that:
$$\forall b(b = 0, a_0 a_1 a_2 \ldots a_m b_{m+1} b_{m+1} \ldots \rightarrow p_n(b) \neq 0).$$
5. In the formulation of *forcing method* in set theory this last condition is expressed as:
$$0, a_0 a_1 a_2 \ldots a_m \Vdash p_n(\dot{x}) \neq 0,$$
where $\dot{x}$ is a *name* for the transcendental real to be constructed.

Now, this situation is used to describe *intuitively* the method of construction of generic models in set theory:

1. We start with transitive submodels $(M, \in)$ of the standard universe $(V, \in)$.

2. We construct minimal submodels, thus similar to the prime field $\mathbb{Q}$.

3. We construct *generic extensions* $M \subseteq N$ by adjoining *generic* sets G, thus corresponding to transcendental numbers: $N = M[G]$.

4. G is describable by infinitely many formulas in the ground model M. It is constructed in a countable recursive process with the use of countably many requirements which can be expressed in M.

Thus a new method created in set theory is explained by a reference to well-known techniques from algebra. It should probably be added that the constructions in question also involve some extra logical or set theoretical results: the Downward Löwenheim-Skolem Theorem gives us countable models and the Mostowski Collapsing Lemma provides transitive models.

Extremal axioms in set theory

Abraham Fraenkel proposed his *Axiom of Restriction* in order to obtain a certain limitation of set theory (Fraenkel 1928). In short, the axiom says that there do not exists any sets except these whose existence can be proven from the axioms of set theory. Of course, this is not an axiom of the object language – it was formulated in the metalanguage. It is an example of *extremal axiom*, in this case an axiom of minimality. Some other logicians (notably Gödel, and later also Myhill and Suszko) also proposed such axioms. However, a few decades later axioms of restriction were strongly criticized as not congruent with the spirit of set theory – cf. Fraenkel, Bar Hillel and Levy 1973. Gödel also took the position that one should rather look for maximality axioms in set theory, playing a similar role to that of the *Axiom of Completeness* in Hilbert's *Grundlagen der Geometrie* (Hilbert 1899). Since then maximal axioms – in the form of axioms of existence of large cardinal numbers – have been intensively investigated. One thing is interesting in this change of opinion, as far as the subject of this article is concerned. This is the fact that the motivation behind the rejection of minimal axioms and acceptation of maximal axioms in set theory is partly of a *pragmatic* character: we simply desire to have as rich a universe of sets as possible. And behind this motivation an *intuitive explanation* is placed: let the universe of set theory be characterized by conditions comparable to the maximality principle (*Axiom of Completeness*) in geometry.

1.5.6 Limits of intuitive explanations

There exist restrictions to intuitive explanations in mathematics. This concerns, among others, several *pathological* objects. As an example, let us consider a few facts concerning infinitely dimensional spaces:

1. In a finite dimensional linear space any two norms are uniformly equivalent. However, this is no longer true in the case of infinite dimensions. For example, if we define two norms in the linear space $C[0,1]$ of all continuous functions from $[0,1]$ to $\mathbb{R}$ by:

$$\|f\|_1 = \max\{|f(t)| : 0 \leqslant t \leqslant 1\}$$

$$\|f\|_2 = \int_0^1 |f(t)|dt,$$

then $(C[0,1], \|\cdot\|_1)$ becomes a complete space, while $(C[0,1], \|\cdot\|_2)$ is not such a space.

2. A subset $A \subset \mathbb{R}^n$ is compact if and only if it is closed and bounded. However, this equivalence is no longer true in the case of infinite dimensions. For example, in the space l^2 (the space of square-summable sequences) with the metric

$$d(\mathbf{x}, \mathbf{y}) = \sqrt{\sum_{i=1}^{\infty} (x_i - y_i)^2}$$

the unit ball

$$B(0,1) = \{\mathbf{x} \in l^2 : d(\mathbf{0}, \mathbf{x}) \leqslant 1\}$$

is closed and bounded but not compact. This is so because the sequence $\{\mathbf{e}_i\}_{i \geqslant 1}$ is included in $B(0,1)$ but $d(\mathbf{e}_i, \mathbf{e}_j) = \sqrt{2}$ for $i \neq j$, and hence this sequence does not contain a convergent subsequence (which in metric spaces is equivalent to compactness).

3. In a finite dimensional linear space (say, $\mathbb{R}^n$) for any convex and bounded set A there exists a point $x_0 \in A$ such that for all $y \in A$: $d(x_0, y) < \operatorname{diam}(A)$. This means that A is included in a ball with center x_0 and radius less than the diameter of A. However, this is no longer true in the case of infinite dimensions. For example, let us define a subset A of the space $(C[0,1], \|\cdot\|_1)$ discussed above as follows:

$$A = \{f \in C[0,1] : 0 = f(0) \leqslant f(t) \leqslant f(1) = 1\}.$$

Then A is convex and bounded, its diameter equals 1, but for every $f \in A$ there exists $g \in A$ such that the distance between f and g equals 1. This means that A is not included in a ball with center belonging to A and radius less than the diameter of A.

In low dimensions as well we encounter situations in which we face clashes with our intuition, or it may be better to say that we find unexpected (not to say unwilling) phenomena. As an example let us mention the fact that $\mathbb{R}^4$ is the only space (among all $\mathbb{R}^n$) in which there exists continuum many mutually not diffeomorphic differential structures. Such exotic structures also exist on the spheres in some dimensions. It may be that it will take several decades (or even more) to get some – as yet unknown – perspective from which we could domesticate such phenomena.

Smale's theorem about eversion of the two-dimensional sphere S^2 in the space $\mathbb{R}^3$ is still called a paradoxical result. Observe that in this case we start with a well-known object (S^2) and the proof uses not very sophisticated tools (homotopy). However, the result clashes with intuitions. What is more interesting is that there are nice visualizations (videos) explaining the whole process of eversion. Should we expect that after internalizing such expositions we will modify our intuitions so that the result is considered completely natural?

The current teaching of mathematics pays a lot of attention to the traditional way of extending the field of rational numbers: to the field of real numbers and then to the field of complex numbers. The topological structures taken into account are based on the absolute value. There also exist different extensions of the field $\mathbb{Q}$, in which the topology is based on another norm, namely p-adic fields with p-adic norm. Here the notion of proximity is related to factorization of the numbers, roughly speaking. The number theory and analysis in the p-adic case is well developed, and there are also many applications of p-adic structures. It is difficult to foresee the future development of tendencies in the teaching of mathematics, but we think it could prove instructive for students to be informed about the possibilities of extending $\mathbb{Q}$ with at least some of the implications that p-adic structures offer. Similar remarks can be made with respect to the hyperreal numbers in the teaching of mathematics. Discussing such topics may deepen the understanding of what is considered to be *natural* in mathematics.

1.6 Closing remarks

I sincerely admit that my experience in teaching mathematics is very modest: it is limited to introductory courses on mathematical logic and an introduction to mathematics addressed to students of the humanities. In addition, I teach a course on mathematical problem-solving to students of cognitive science. I have not conducted any research in the domain of education science.

However, due to the fact that my audience comprises students of the humanities, I have always been interested in ways, methods, and tools which could be helpful in the process of such students understanding mathematical ideas. They are often suspicious of mathematics, have bad memories from school, they do not believe that they can ever be successful with mathematics. In extreme cases, are even claim to be proud of being unable to understand mathematical ideas.

Literature describes several methods or even tricks used to make sophisticated ideas more accessible to students, for instance to familiarize them with the concepts of infinity, continuity, derivative, integral, or some more advanced arithmetical, algebraic or topological notions. The efficiency of such methods can be measured by the ability of students to solve problems (of a certain kind) by themselves. We think that at least two issues are interesting in this respect. First, is a given method purely local in character, or can it be applied to the situation where the student meets a completely new kind of problem, not explained intuitively before? For example, can they apply a kinematic explanation of derivative to, say, variational problems? Secondly, what are the historical changes in the style of intuitive explanations? For example, does students' familiarity with computers (computer programs) influence the style of modern explanations?

It is a sad and to a certain degree mysterious fact that most of the population has difficulty understanding of mathematics. Poincaré wrote:

> How is it that there are so many minds that are incapable of understanding mathematics? Is there not something paradoxical in this? Here is a science which appeals only to the fundamental principles of logic, to the principle of contradiction, for instance, to what forms, so to speak, the skeleton of our understanding, to what we could not be deprived of without ceasing to think, and yet there are people who find it obscure, and actually are the majority. That they should be incapable of discovery we can understand, but that they should fail to understand the demonstrations expounded to them, that they should remain blind when they are shown a light that seems to us to shine with a pure brilliance, it is altogether miraculous.
>
> And yet one need have no great experience of examinations to know that these blind people are by no means exceptional beings. We have here a problem that is not easy of solution, but yet must engage the attention of all who wish to devote themselves to education. (*Science and Method*, citing Sierpińska 1994, 112)

The present paper is only an introductory note. Let us close with examples of research problems we expect to face:

1. Are intuitive explanations of a purely local character (in the sense that an explanation is related to one given problem only, and cannot be repeated in other cases)?
2. How effective are they? How can this effectiveness be measured? Are teachers' reports sufficient and reliable data?
3. How can one overcome the dangers of oversimplification and misunderstanding which can arise in the context of intuitive explanations?
4. How can one to be sure of not cheating with intuitive explanations?
5. To what extent is empirical research in teaching possible and ethically acceptable?
6. What kinds of experiments could be proposed? For example, do the following projects seem promising:
   (a) Use of eye-tracking of perception of impossible figures.
   (b) Making videos explaining concepts of advanced mathematics instructive for our kind of audience (adult students in the humanities, and in particular in cognitive science).
   (c) Designing mathematical games, in particular computer games, for the same purposes as mentioned above.
   (d) Drawing up tests checking understanding of geometric transformations and diagrams.
   (e) Drawing up tests evaluating the development of mathematical intuitions in adults.

Bibliography

Aberdein, A., Dove, I.J. (Eds.) 2013. *The argument of mathematics*. Springer Science+Business Media, Dordrecht.

Adams, C.C. 2004. *The knot book. An elementary introduction to the mathematical theory of knots*. American Mathematical Society, Providence, Rhode Island.

Ajdukiewicz, K. 1974. *Pragmatic logic*. D. Reidel Publishing Co., Dordrecht Boston, Polish Scientific Publishers, Warszawa.

Benacerraf, P., Putnam, H. 1964. *Philosophy of mathematics: selected readings.* Prentice-Hall, Englewood Cliffs, N.J.

Ben-Zeev, T., Star, J. 2001. Intuitive mathematics: theoretical and educational implications. In: Torff, Sternberg (Eds.) 2001, 29–56.

Byers, W. 2007. *How mathematicians think. Using ambiguity, contradiction and paradox to create mathematics.* Princeton University Press, Princeton and Toronto.

Conway, J.H., Guy, R.K. 1998. *The book of numbers.* Springer Verlag, New York.

Davis, P.J., Hersh, R. 1981. *Mathematical experience.* Birkhäuser, Boston.

Feferman, S., Friedman, H.M., Maddy, P., Steel, J.R. 2000. Does mathematics need new axioms? *The Bulletin of Symbolic Logic* 6 (4), 401–446.

Fischbein, H. 1987. *Intuition in science and mathematics: an educational approach.* Kluwer Academic Publishers, New York / Boston / Dordrecht / London / Moscow.

Fraenkel, A.A. 1928. *Einleitung in die Mengenlehre.* Verlag von Julius Springer, Berlin.

Fraenkel, A.A., Bar-Hillel, Y., Levy, A. 1973. *Foundations of set theory.* North-Holland Publishing Company, Amsterdam London.

Galperin, G. 2003. Playing pool with π (The number π from a billiard point of view). *Regular and Chaotic Dynamics* 8 (4), 375–394.

Ghrist, R. 2014. *Elementary applied topology.* Createspace, ISBN 978-1502880857.

Gödel, K. 1986–2003. S. Feferman *et al.* (Eds.) *Kurt Gödel: Collected Works, Volume I 1986, Volume II 1990, Volume III 1995, Volume IV 2003, Volume V, 2003.* Oxford University Press, New York.

Grzegorczyk, A. 1962. Uzasadnianie aksjomatów teorii matematycznych. *Studia Logica* 13, 197–202.

Hadamard, J. 1954. *An essay on the psychology of invention in the mathematical field.* Dover Publications, Inc., New York.

Hanna, G., Jahnke, H.N., Pulte, H. (Eds.) 2010. *Explanation and proof in mathematics. Philosophical and educational perspectives.* Springer, New York Dordrecht Heidelberg London.

Havil, J. 2007. *Nonplussed! Mathematical proof of implausible ideas*. Princeton University Press, Princeton and Oxford.

Heath, T.L. (Ed.) 2002. *The works of Archimedes*. Dover Publications, Inc., Mineola, New York.

Hilbert, D. 1899. *Grundlagen der Geometrie*. Festschrift zur Feier der Enthüllung des Gauss-Weber-Denkmals in Göttingen. Teubner, Leipzig.

Hilbert, D., Cohn-Vossen, S. 1932. *Anschaunliche Geometrie*. Verlag von Julius Springer, Berlin.

Kasner, E., Newman, J. 1940. *Mathematics and the imagination*. Simon and Schuster, New York.

Kitcher, P. 1984. *The nature of mathematical knowledge*. Oxford University Press, Oxford.

Klymchuk, S., Staples, S. 2013. *Paradoxes and sophisms in calculus*. Mathematical Association of America, Washington, DC.

Koepke, P. 2013. Models of set theory. Available at:

http://www.math.uni-bonn.de/ag/logik/teaching/2013SS/models_of_set_theory_1/Skript.pdf

Krygowska, Z. 1967. Rozumowanie empiryczne, intuicyjne i formalne w nauczaniu matematyki. *Roczniki Polskiego Towarzystwa Matematycznego*, series II: *Wiadomości Matematyczne* 10, 49–91.

Lakatos, I. 1976. *Proofs and refutations. The logic of mathematical discovery*. Cambridge.

Lange, M. 2014. Depth and explanation in mathematics. *Philosophia Mathematica* Advance Access published September 12, 2014:

http://philmat.oxfordjournals.org/

Laraudogoitia, J.P. 1996. A beautiful supertask. *Mind* 105, 81–83.

Lénárt, I. 1996. *Non-Euclidean adventures on the Lénárt sphere*. Key Curriculum Press, USA.

Levi, M. 2009. *The mathematical mechanic. Using physical reasoning to solve problems*. Princeton University Press, Princeton and Oxford.

MacLane, S. 1986. *Mathematics. Form and function*. Springer Verlag, New York Berlin Heidelberg Tokyo.

Mancosu, P. 2015. Explanation in mathematics. *The Stanford Encyclopedia of Philosophy* (Summer 2015 Edition), Edward N. Zalta (ed.), URL = http://plato.stanford.edu/archives/sum2015/entries/mathematics-explanation/

Needham, T. 1997. *Visual complex analysis*. Clarendon Press, Oxford.

Nickerson, R.S. 2010. *Mathematical reasoning. Patterns, problems, conjectures, and proofs*. Psychology Press, New York London.

Parsons, C. 2008. *Mathematical thought and its objects*. Cambridge University Press, Cambridge, New York, Melbourne, Madrid, Cape Town, Singapore, São Paulo.

Pogonowski, J. 2016. Kontekst przekazu w matematyce. *Annales Universitatis Paedagogicae Cracoviensis. Studia ad Didacticam Mathematicae Pertinentia* 8, 119–137.

Pogonowski, J. 2018. Intuitive explanations of mathematical ideas. *Annales Universitatis Paedagogicae Cracoviensis. Studia ad Didacticam Mathematicae Pertinentia* 10, 123–137.

Pogonowski, J. 2019. *Extremal axioms. Logical, mathematical and cognitive aspects*. Wydawnictwo Nauk Społecznych i Humanistycznych UAM, Poznań.

Polya, G. 2009. *Mathematical discovery on understanding, learning, and teaching problem solving*. Ishi Press International, New York, Tokyo.

Polya, G. 2014. *Mathematics and plausible reasoning*. Vol.I: *Induction and analogy in mathematics*, Vol. II: *Patterns of plausible inference*. Martino Publishing, Mansfield Centre, CT.

Prasolov, V.V. 2011. *Intuitive topology*. American Mathematical Society.

Romero, G.E. 2014. The collapse of supertasks. *Foundation of science* 19 (2), 209–216.

Schoenfeld, A. H. 1985. *Mathematical problem solving*. Academic Press, Inc., Orlando.

Schoenfeld, A.H. 1992. Learning to think mathematically: Problem solving, metacognition, and sense-making in mathematics. In: D. Grouws (Ed.), *Handbook for research on mathematics teaching and learning*. MacMillan, New York, 334–370.

Scholten, P., Simoson, A. 1996. The falling ladder paradox, *The College Mathematics Journal* 27 (1), 49–54.

Sierpińska, A. 1994. *Understanding in mathematics*. The Falmer Press, London.

Steiner, M. 1978. Mathematical explanation. *Philosophical Studies: An International Journal for Philosophy in the Analytic Tradition* 34 (2), 135–151.

Tall, D. 2013. *How humans learn to think mathematically. Exploring the three worlds of mathematics*. Cambridge University Press, Cambridge.

Tao, T. 2009. There's more to mathematics than rigour and proofs. Available at:

https://terrytao.wordpress.com/career-advice/theres-more-to-mathematics-than-rigour-and-proofs/

Tieszen, R.L. 1989. *Mathematical intuition: phenomenology and mathematical knowledge*. Kluwer Academic Publishers, Dordrecht.

Torff, B., Sternberg, R.J. (Eds.) 2001. *Understanding and teaching the intuitive mind: student and teacher learning*. Lawrence Erlbaum Associates Publishers, Mahwah New Jersey London.

Thurston, W. 1994. On proof and progress in mathematics. *Bulletin of the American Mathematical Society* 30 (2), 161–177.

Chapter 2

A note on famous mistakes in mathematics

2.1 Hunting for mistakes

Most mistakes by mathematicians remain hidden, forgotten, and drowned in oblivion. They are seldom recalled by the authors of mathematical publications. However, it happens sometimes that a mistake becomes famous and gives rise to new mathematical ideas.

The currently accepted style of mathematical publications is partly responsible for the difficulties in investigating mistakes. One may call it *the Gauss style*: only the final result is shown in a publication, without any hint as to how it was obtained, without any overt discussion of the *context of discovery*. Some of Euler's publications of are written in a different style, though: the author discusses the path he took to the final result, mentioning the difficulties encountered along the way. We believe that this style, which one might call *the Euler style*, could be appropriate in teaching mathematics.

A few terminological remarks are in order. The concepts we are interested in include errors, lapses, flaws, mistakes, fallacies, contradictions, ambiguities, wrong assumptions, false analogies, hasty generalizations, incomplete proofs, lack of entailment, and even correct results formerly rejected by the mathematical community. Thus we are taking a rather broad approach to the problem.

A question that naturally arises is whether it is at all possible to say anything interesting in such a general approach. Other options would be to concentrate on a given period in mathematics or on a particular type of mistake (say, false analogy). Beyond our interest here are the following: paradox, sophism, fraud, puzzle, conundrum, brain-teaser, and howlers. Note that many sophisms were created purposively, in order to warn readers about the pitfalls of relying on illusory intuitions.

The term *cognitive bias* refers (in cognitive psychology) to a systematic pattern of deviation from norm or rationality in judgment. We speak of cognitive biases mostly in connection with a subjective interpretation (or representation) of the – presumably objective – social world. Thus, cognitive biases are investigated mainly from the perspective of human judgments, decision-making and economic behavior. Since recently it has been well known that, on average, people do not reason rationally, they have problems with calculations (innumeracy), they take irrelevant information as the base for decision-making, they jump to conclusions, they are driven by emotions, they strongly believe themselves to be "intuitive statisticians" correctly estimating the chances of an event without calculating the probabilities, and so on.

Here we are interested in the issue whether the concept *cognitive bias* is applicable to mathematics. Is it possible to follow "a systematic pattern of deviation from norm or rationality in judgment" in mathematics? This would imply that mathematics is – to a certain degree – a normative discipline. There is also the possibility that mistakes in mathematics should be conceived of in a different manner than the cognitive biases mentioned above. There is then naturally the question how it should be done.

Is mathematical reality a part of social reality? This question is of course related to the old discovery-invention dilemma, the question of whether mathematics was invented or discovered. Each solution to this dilemma leads to further questions, for instance how to distinguish a *mistake in invention* from a *mistake in discovery*.

One should stress the difference between the meaning of *proof* in logic and in mathematics. From the point of view of logic a proof is a precisely determined syntactic construction, consisting of formulas interrelated by the logical entailment. On the other hand, a mathematical proof bears a pragmatic mark: its acceptance is relativized to the opinion of the mathematical community. Hence mistakes in mathematics may be caused not only by the lack of entailment but also by other factors.

Mathematical mistakes are relativized to a given epoch in the same way as the criteria of correctness were accepted in that epoch. Of special interest are the circumstances of mistakes. They may include, it seems, the influence of tradition, illusory suggestions from the empirical realm, and sometimes even a wishful thinking.

Finally, let us add that we do not count as mathematical mistakes those mathematical models of nature which were later falsified. For example, Kepler's geometrical model of the (then known) Solar system is not a *mathematical* mistake. It was later replaced by a more adequate model.

2.2 Types of mistakes

It is impossible to give any reasonably complete and adequate classification of mathematical mistakes – the ways in which errors are committed are innumerable and cannot be predicted in advance. However, it is justified to discuss the role of mistakes in the development of mathematical knowledge (compare the well-known dictum *erro ergo disco*). Let us observe that it also happens that the mathematical community classifies as mistaken some views which later are recognized as perfectly correct.

Mathematical mistakes are described and analyzed in several works. The text of Euclid's *Pseudaria*, mentioned by Proclus is lost to us. Presumably, it contained several examples of false proofs (i.e. sophisms) presented as warnings for students of geometry. The issue is discussed in some detail in Acerbi 2008. *A Budget of Paradoxes* written by Augustus De Morgan includes descriptions of numerous mistakes (De Morgan 1915). Maurice Lecat collected about 500 mistakes made by 300 famous mathematicians in Lecat 1935. Furthermore, some more recent monographs include discussion of mistakes made by famous mathematicians (for instance Posamentier, Lehmann 2013). But usually the discussion of mistakes is limited to errors made by students (see for instance Barbeau 2000, Bradis, Minkovskii and Kharcheva 1999, Bunch 2014, and Maxwell 1959). Mistakes, incomplete proofs and the like are also discussed in the internet.

It is worth mentioning that there are also collections of counterexamples in different domains of mathematics. They do not list mistakes, but nevertheless point to the dangers of committing an error – see for instance Gelbaum and Olmsted 2003, Steen and Seebach 1995, or Wise and Hall 1993.

What could be the possible causes of mathematical mistakes? We think that at least the following could be listed: incompetence, oversight, illusory physical intuitions, suggestive drawings, verbal problems, wrong conceptualizations, lack of sound logical foundations or the complexity of the notions under investigation.

A tentative typology of mathematical mistakes might take the following form: material mistakes (wrong assumptions), formal mistakes (non sequitur), incomplete proofs, errors in calculations, improper inductive hints, false analogies, wrong generalizations, false conjectures (though strongly believed at the moment of their formulation), collapsed research programs, and so on. We wish to emphasize once again that it is not possible to predict particular mistakes and their circumstances. Nevertheless, we can attempt to put together catalogues of "typical" (genuine) mistakes.

2.3 Examples of mistakes

Well known classic examples include: fruitless efforts in proving the 5th postulate in Euclidean geometry, false proofs of Fermat's theorem, several deviant proofs based on hidden intuitions (such as drawings or kinematic hints), assumptions leading to antinomies (for example Russell's antinomy), and absurd results in the summation of infinite series (before the appropriate conditions of convergency were set up), and so on. We will discuss some of these in more detail, paying attention to a few mistakes made by prominent logicians as well. Several developments in the history of mathematics are considered revolutionary, for instance the discovery of incommensurable magnitudes, the invention of analytic geometry, the origins of Calculus, or the structural revolution in the 19th century. Such changes in the paradigm of mathematical research do not necessarily imply that previous investigations were faulty. Rather, they show that mathematics gained a new perspective. Some of the examples mentioned below are discussed in Posamentier and Lehmann 2013. We also referred to them in Pogonowski 2019, 2020.

Euclid's 5th postulate. It took a long time to show the independence of this postulate from the other ones. At the beginning, there were numerous attempts to prove it on the basis of the remaining postulates. They all failed, but nevertheless some important observations were made during those years. Thus, for example:

Proclus (410–485) wrote his *Commentary* on false proofs (of the postulate) and then gave his own proof, false as well. *Alhazen* (965–1040) introduced what was to become known as the *Lambert quadrilateral* (a quadrilateral with three right angles) and used concepts related to motion in geometry in his attempts to prove the postulate by reductio ad absurdum. He also obtained some theorems of hyperbolic and elliptic geometry. *Omar Khayyam* (1048–1131) was the first not to commit *petitio principii* positing that the postulate be inferred from a more intuitive one. He did not accept motion in geometry, introduced what came to be known as the *Saccheri quadrilateral* (a quadrilateral with two equal sides perpendicular to the base), and recognized three possibilities obtained by omitting the fifth postulate. *Nasir al-Din Tusi* (1201–1274) tried to prove the postulate by applying apagogic method, and mentioned (but rejected) hyperbolic and elliptic geometries. *Giordano Vitale* (1633–1711) observed that a line which at all points is at equal distance from a given straight line must itself be a straight line. *Giro-*

lamo Saccheri (1667–1733) tried a reductio ad absurdum proof. Since the fifth postulate is equivalent to the statement that the sum of angles of a triangle equals π, he considered two cases: when the sum is smaller than π and when it is bigger than π. As a consequence of the first case, straight lines would be limited, which contradicted Euclid (but note that this is exactly the case in elliptic geometry). In the second case Saccheri obtained several consequences, all of which he felt contradicted intuition. One such consequence was the existence of a triangle with maximum area (and this is exactly the case in hyperbolic geometry). *Johann Heinrich Lambert* (1728–1777) made use of the Lambert quadrilateral and eliminated the possibility of its fourth angle being obtuse. He proved several theorems under the assumption that this angle was acute – among others that the sum of angles of a triangle is growing when the area of the triangle is diminishing. He did not think that these results contradicted intuitions, and even speculated about possible models. *Carl Friedrich Gauss* (1777–1855) considered a system of geometry with the negation of the fifth postulate, but did not publish his considerations. In 1829 *Nicolai Lobachevsky* (1792–1856) has published a work in which the fourth angle in the Lambert quadrilateral was acute. *János Bolyai* (1802–1860) independently of Lobachevsky published a paper in 1831 concerning the same system of geometry. Lobachevsky, Bernhard Riemann (1826–1866) and Henri Poincaré (1854–1912) developed elliptic and hyperbolic geometry. *Eugenio Beltrami* (1835-1899) proved in 1868 the independence of the 5th postulate from the other postulates of the Euclidean system. Thus we can see that the path to the final solution of a mathematical problem may be very twisted and saturated with several types of mistakes.

Galileo: brachistochrone. Galileo wrongly claimed that the brachistochrone (that is, the curve of fastest descent) is a segment of a circle and not any succession of line segments. One may suspect that the claim was formulated after a series of mechanical experiments. However, the curve in question is not a fragment of a circle but a fragment of the *cycloid*. The investigation of such curves gave impulse to the origin of the calculus of variations.

Origins of calculus, infinitesimals. It is well known and fully described in literature that the origins of calculus lacked sound logical foundations. This concerns Newton as well as Leibniz, although they referred to different intuitions (kinematic and geometric in the case of Newton, geometric and algebraic in the case of Leibniz). Newton's approach obtained a solid mathematical ground in the 19th century (arithmetization of analysis), while the approach proposed by Leibniz found a precise mathematical formulation in non-standard analysis in the 20th century. As we can see, mathematicians played with calculus, infinite series, and infinitesimals decade after decade, with successful results, although they were at the same time aware of logical inconsistences.

Infinite series: Grandi series, Euler proofs. Mathematicians in the 17th and 18th centuries were working without clear and precise criteria for convergence and sometimes obtained curious results. Let us mention that Euler, when solving the Basel Problem, used an argument from analogy (writing about infinite products involving trigonometric functions). Only later did Weierstrass prove a factorization theorem justifying Euler's argument.

Solutions by radicals: naive intuitions abolished. Many attempts at formulating a solution by radicals of a polynomial of an arbitrary degree have probably been forgotten. Having such solutions for second, third and fourth degree might obviously provide (an illusory!) hint that the problem is solvable for any degree. Those hopes were dashed by the works of Galois, as is well known.

Fermat's theorem: several failures. Wiles' result finally proved Fermat's Last Theorem. However, partly because the formulation of the theorem was so simple, it attracted many mathematical cranks who tried to prove it either by brute force or by very elementary arithmetic reasonings. Its flawed proof by Lamé used an unjustified assumption about the uniqueness of factorization in some algebraic structures. Let us add that the postulate of existence of ideal numbers (Kummer) is an example showing how mathematical ideology can lead to new discoveries. That certain algebraic structures do not have the property of unique decomposition could be conceived as undesirable, but it is a fact. However, the property of unique factorization is of such importance that it was worth postulating the existence of completely new mathematical objects: Kummer's ideal numbers (later generalized by Dedekind in his theory of ideals).

Early rejection of negative and imaginary numbers. These kinds of objects, suspicious at the time of their origin, appeared in mathematics early as unexpected, unwilling intruders in calculations. They obtained names emphasizing their peculiarity (such as imaginary, false, dumb). But they were necessary in order to conduct calculations. In both cases – negative as well as imaginary numbers – it took a few centuries before such numbers became fully domesticated. They began functioning as legitimate mathematical objects after obtaining general results proving that they formed a sound system in which one was able to perform prescribed operations.

Cauchy: continuity and infinitesimals. There is still lively discussion concerning the matter of whether Cauchy was right or wrong in claiming that a limit of a (pointwise) convergent sequence of continuous functions was itself continuous. On the one hand, it is claimed that he was wrong: a stronger assumption of uniform convergence is necessary (within classical analysis).

On the other hand, it is important to remember that Cauchy made essential use of the concept of an infinitesimal, hence he could have understood continuity differently than in the classical concept.

Lebesgue: projections of Borel sets. Lebesgue wrote that projections of Borel sets are Borel, a mistake that was corrected by Luzin. This gave rise to a whole domain of descriptive set theory dealing with projective sets.

Poincaré: the three-body problem. According to anecdote, Poincaré withdrew his article on this topic when he noticed a mistake in it. However, the mistake turned out to be fruitful, leading to later work concerning chaos theory.

Number theory: several false conjectures. Numbers of the form $F_n = 2^{2^n} + 1$ are called Fermat numbers. Fermat claimed that all numbers of this form are prime. However, before long Euler revealed F_5 to be a composite number:

$$F_5 = 2^{32} + 1 = 4294967297 = 641 \cdot 6700417.$$

Euler claimed that the equations:

$$a^3 + b^3 = c^3,$$

$$a^4 + b^4 + c^4 = d^4,$$

$$a^5 + b^5 + c^5 + d^5 = e^5,$$

and so on, do not have solutions in natural numbers. We know that counterexamples to this claim can be found for sums of fourth and sums of fifth powers. Moreover, there are infinitely many such counterexamples in the case of fourth powers.

Four colors theorem. The hypothesis that four colors are sufficient for coloring any map on a plane was formulated in 1840 by August Möbius (and independently by Francis Guthrie in 1852). Then there followed several unsuccessful attempts towards a proof of this hypothesis (Alfred Kempe in 1879, Peter Guthrie Tait in 1880). The hypothesis was finally proven in 1976 by Kenneth Appel and Wolfgang Haken. It is worth recalling that this was the first proof in which the usage of a computer program proved essential, which of course provoked a discussion on whether such proofs are *genuine* proofs.

Nikolai Chebotaryov: factorization of $x^n - 1$. Let us look at the first few lines of a table showing the factorization of the polynomial in question:

n	$x^n - 1$	Decomposition
1	$x^1 - 1$	$x - 1$
2	$x^2 - 1$	$(x-1)(x+1)$
3	$x^3 - 1$	$(x-1)(x^2+x+1)$
4	$x^4 - 1$	$(x-1)(x+1)(x^2+1)$
5	$x^5 - 1$	$(x-1)(x^4+x^3+x^2+x+1)$
6	$x^6 - 1$	$(x-1)(x+1)(x^2+x+1)(x^2-x+1)$
7	$x^7 - 1$	$(x-1)(x^6+x^5+x^4+x^3+x^2+x+1)$
8	$x^8 - 1$	$(x-1)(x+1)(x^2+1)(x^4+1)$
9	$x^9 - 1$	$(x-1)(x^2+x+1)(x^6+x^3+x^2+x+1)$
10	$x^{10} - 1$	$(x-1)(x+1)(x^4+x^3+x^2+x+1)(x^4-x^3+x^2-x+1)$

Chebotaryov claimed that all coefficients in such factorizations follow this pattern (i.e. are equal 1 or -1). However, the pattern collapses for $n = 105$, where the coefficient of the seventh power (as well as the forty first power) equals -2. More counterexamples are known.

Malfatti circles. Gian Francesco Malfatti (1731–1807) formulated the problem of carving three circular columns out of a triangular block of marble, using as much of the marble as possible. The problem can of course be reduced to a planimetric level (concerning the bases of the pillars). Malfatti conjectured that three mutually-tangent circles inscribed within the triangle would provide the optimal solution. Such circles are called *Malfatti circles.* His conjecture was wrong, which was shown no sooner than in 1930.

The Mertens conjecture. Mertens claimed that $|M(n)| < \sqrt{n}$ for all n, where $M(x) = \sum_{y \leqslant x} \mu(y)$ and μ is the Möbius function. The falsity of this conjecture was shown in Odlyzko and te Riele 1985. However, the first counterexample is bigger than 10^{14}. We also know the upper bound for the first such counterexample: $\exp(1.59 \times 10^{40})$. Properties of the function $M(n)$ are connected with the Riemann Hypothesis. A discussion regarding what it could mean that the Riemann Hypothesis is *true with probability one* may be found, for example, in Davis and Hersh 1981.

The Borsuk conjecture. Karol Borsuk demonstrated that a three-dimensional ball can be decomposed into four solids each of which has a diameter smaller than the initial ball (Borsuk 1933). He then posed the following problem: *Can every bounded subset X of the space $\mathbb{R}^n$ be partitioned into $(n+1)$ sets, each of which has a smaller diameter than X?* Several partial positive results were obtained, thus suggesting that the answer may be positive in a general case. However, the answer turned out to be negative (Kahn and Kalai 1993). The first counterexamples were found in rather high dimensions.

The Perko pair. This is an example of mistake due to oversight. Two rather complicated knots (denotebd by 10_{161} and 10_{162} in Dale Rolfsen's knot table) were later shown to be essentially the same knot.

Lion and man problem. This problem was posed by Rado in 1925, and concerns a game of pursuit. A man and a lion have identical maximum speed, and the lion tries to catch the man. They are situated on an island of the shape of a circle. Who has a winning strategy? For a few decades it was accepted that the lion has a winning strategy in this game. However, using a result of Besicovitch it can be shown that it is the man who has a winning strategy, though the lion becomes infinitely close to the man (as always, we make the ridiculous assumption that both the lion and the man are points). The solution is based on the fact that the harmonic series is divergent.

Several misfortunes in mathematical performance are described in the books by Underwood Dudley (for instance Dudley 1992). There still exist *mathematical cranks (Trisectors, Fermatists, and Goldbachers, and so on,* in Dudley's terminology) who try to prove the Goldbach Hypothesis or to show that it is possible to trisect an angle with straight edge and compass. The same concerns many flawed attempts of the quadrature of the circle using ruler and compass alone, even though it is well known that it is impossible.

As it happens, recognized mathematicians also sometimes promote very bizarre opinions. Louis de Branges is a well-known mathematician (he proved the Bieberbach conjecture). He has claimed to have a proof of the Riemann hypothesis, but as far as we know it has not been accepted by the mathematical community. Wolfgang Mückenheim presents several arguments to show that non-denumerable sets are fictions (Mückenheim 2011), but these are also criticized by other mathematicians.

Siberian "proof" of the FLT. The internet page

http://www.sciteclibrary.ru/eng/catalog/pages/6253.html

contains an alleged proof of the Fermat's Last Theorem based on a ridiculous probabilistic-like argument. The argument runs as follows. FLT states that the equation

$$x^n + y^n = z^n,$$

where $z \neq 0$, $x < z$, $y < z$, $n \geqslant 3$ does not have solutions in natural numbers x, y, z. Let us divide both sides of the equation by z^n. We get:

$$(\frac{x}{z})^n + (\frac{y}{z})^n = 1.$$

Now, let us think (sic!) about events A and B with the following probabilities: $P(A) = (\frac{x}{z})^n$ and $P(B) = (\frac{y}{z})^n$. Let also $P(A \cup B) = 1$. Then $P((A \cup B)^c) = 0$ (where X^c denotes the event complementary to X). By De Morgan Laws we have:

$$0 = P((A \cup B)^c) = P(A^c \cap B^c) = P(A^c) \cdot P(B^c) = (1 - (\frac{x}{z})^n) \cdot (1 - (\frac{y}{z})^n).$$

Hence either $x = z$ or $y = z$. The initial equation has thus only trivial solution and therefore the FLT is proved. The argument is discussed, with a few comments, by Alexander Bogomolny on his page http://www.cut-the-knot.org/ctk/ErrDisc.shtml The author of the discussed argument speculates that perhaps Fermat himself had such an idea in mind and bravely suggests that the argument shows a contradiction between axioms of arithmetic and axioms of probability theory. We have discussed this example in Pogonowski 2019.

Let us also mention a few examples of mistakes from the history of logic:

Gottlob Frege: unrestricted comprehension axiom. The assumption that any property can determine a set leads to antinomy (Russell's antinomy). One can avoid a contradiction accepting a suitable restriction of the comprehension axiom. This issue is described in full detail in many places.

Lewis Carroll: heuristic rules involving resolution. Lewis Carroll suggested a method of resolving soriteses based on a kind of a resolution rule. However, the rule did not work in a general case. He had noticed that and proposed a completely new approach: the method he termed *The Method of Trees*. This was actually a prototype of the analytic tableaux method developed half a century later. We have discussed this issue in Pogonowski 2008.

Ernst Zermelo and Skolem's Paradox: a flawed proof. Zermelo was deeply disappointed by the fact that his invention, axiomatic set theory, could have a denumerable model. He tried to show that Skolem's Paradox is not a paradox at all. However, the proof he gave was flawed, in the sense that he implicitly assumed (without justification) that models of set theory should be closed with respect to arbitrary unions (see van Dalen and Ebbinghaus 2000).

Ernst Zermelo and Kurt Gödel: misinterpretation. The correspondence between Zermelo and Gödel contains a critique of Gödel's proof of the incompleteness theorem. However, the critique is incorrect because Zermelo made an error in confusing object language with metalanguage. More importantly, the correspondence in question shows differences between Zermelo and Gödel as far as the belief in ultimate decidability of all mathematical problems is concerned. We have discussed this issue in Pogonowski 2006.

Rudolf Carnap: Gabelbarkeitssatz. Carnap tried to show that the concepts of *categoricity* and *completeness* have the same extension. The proof, formulated in the theory of types, was flawed (see Awodey and Carus 2001). Obviously, categoricity implies completeness but not vice versa. Lindenbaum and Tarski provided a sufficient condition for that vice-versa implication (Lindenbaum and Tarski 1936). It should be noted that such problems have quite recently undergone intensive investigation (the Carnap-Fraenkel property).

Abraham Fraenkel: axiom of restriction. Fraenkel proposed an axiom of restriction that should guarantee the existence of only such sets whose existence could be proved in set theory. The axiom was later criticized, partly on purely pragmatic grounds. Recently the argument has rather been for several kinds of *maximality* axioms in set theory (axioms of existence of large cardinal numbers). Observe that Gödel's axiom of constructibility is another example of an axiom of restriction. We have discussed this issue in Pogonowski 2019.

David Hilbert: impossibility of the consistency proof. Hilbert's ambitious program cannot be realized in its entirety. Thus the belief that one could prove consistency of rich mathematical theories within the theories themselves was mistaken. However, Hilbert's program can be realized partially and its very formulation influenced the development of proof theory.

2.4 Consequences for the epistemology of mathematics

One of the most vividly discussed problems in the philosophy of mathematics is related to the status of mathematical knowledge. Should we talk of *discoveries* of mathematical truths or rather about mathematics as a kind of creative activity, as an *invention* only? We dare claim that the real significance of the controversy concerning this dilemma lies exclusively in the quality of arguments proposed in order to explicate the problem. An ultimate solution to the dilemma is unlikely to be achieved. The spectrum of such arguments embraces those that refer to the "unexplicable" applicability of mathematics in the description of physical reality or to the indispensability of mathematics in science, as well as those attempting to ground the origin and function of mathematics on the sole base of formation of conceptual metaphors. Meanwhile, the purely mathematical results obtained by those who believe in the discovery of mathematical truths and those who believe them to be solely inventions of the human intellect are comparable in depth, quality and formal sophistication.

In addition we have instances of colliding mathematical intuitions. One could mention the differences between Newton's approach to Calculus and that suggested by Leibniz, for example. Newton referred to finite magnitudes only, and relied on hints from geometry and kinematics. Leibniz proposed that infinitely small be entities treated algebraically. An example of a more recent collision in intuitions could be the tension between the axiom of choice and the axiom of determinacy in set theory. Both have a solid mathematical background, but they are mutually inconsistent. However, the axiom of choice has a more universal character. But still, there is an important subfield of set theory where one cannot simultaneously accept both of them.

One may ask whether mathematicians act so as to unify their intuitions? What are the decisive factors responsible for such unification? And one could ask about the possible (and plausible) reasons behind deciding which way mathematics should take in the case of undecidable sentences, of which there are many, both in arithmetic and set theory.

What is the role of mathematical mistakes as far as the ideological disputes in mathematics are concerned? Does intuitionism see mistakes in a different way from that typical for classical mathematics? The same question may be asked with respect to more recent views in the philosophy of mathematics.

Let us observe an important difference between mathematics and the empirical sciences. When a mathematical model applied to physical phenomena appears inadequate, we change the mathematical model but we do not throw it out from mathematics itself. Change in the representation of physical phenomena is what takes place in such a situation.

As far as mathematical practice is concerned we opt for a kind of *mathematical agnosticism*. We think that mathematicians conduct their everyday work in the same manner, regardless of the solution to the discovery-invention dilemma. The philosophical assumptions, and their favorite ideological standpoints, do not seem to influence their work.

We are aware that this brief note has not offered deep insight into the issue of mistakes in mathematics. We have presented only a few examples and commentaries. Instead of general reflections on mathematical mistakes, attention should probably focus on particular mistakes and their circumstances as well as on their impact on the development of mathematics.

Bibliography

Acerbi, F. 2008. Euclid's Pseudaria. *Archive for History of Exact Sciences* 62 (5), 511–551.

Awodey, S., Carus, A.W. 2001. Carnap, completeness, and categoricity: the *Gabelbarkeitssatz* of 1928. *Erkenntniss* 54, 145–172.

Barbeau, E.J. 2000. *Mathematical fallacies, flaws, and flimflam*. the Mathematical Association of America, Washington, DC.

Borsuk, K. 1933. Drei Sätze über die n-dimensionale euklidische Sphäre, *Fundamenta Mathematicae* 20, 177–190

Bradis, V.M., Minkovskii, V.L., Kharcheva, A.K. 1999. *Lapses in mathematical reasoning*. Dover Publications, Mineola, New York.

Bunch, B. 2014. *Mathematical fallacies and paradoxes*. Dover Publications, Mineola, New York.

Byers, W. 2007. *How mathematicians think. Using ambiguity, contradiction and paradox to create mathematics*. Princeton University Press, Princeton and Toronto.

Dalen, D. van, Ebbinghaus, H.D. 2000. Zermelo and the Skolem paradox. *The Bulletin of Symbolic Logic* 6 (2), 145–161.

Davis, J.P., Hersh, R. 1981. *The mathematical experience*. Birkhäuser, Boston.

De Morgan, A. 1915. *A budget of paradoxes*. Volume I. The Open Court Publishing Co., Chicago London.

Dudley, U. 1992. *Mathematical cranks*. The Mathematical Association of America, Washington, DC.

Gelbaum, B.R., Olmsted, J.M.H. 2003. *Counterexamples in analysis*. Dover Publications, Inc., Mineola, New York.

Kahn, J., Kalai, G. 1993. A counterexample to Borsuk's conjecture. *Bulletin of the American Mathematical Society* 29, 60–62.

Lecat, M. 1935. *Erreurs de mathematiciens des origines à nos jours*. Castaigne, Brüssel.

Lindenbaum, A., Tarski, A. 1936. Über die Beschränkheit der Ausdruckmittel deduktiver Theorien. *Ergebnisse eines mathematischen Kolloquiums* 1934–1935, 7, 15–22.

Maxwell, E.A. 1959. *Fallacies in mathematics*. Cambridge University Press, Cambridge.

Mückenheim, W. 2011. *Mathematik für die ersten Semester*. R. Oldenbourg Verlag, München.

Odlyzko, A. M., te Riele, H. J. J. 1985. Disproof of the Mertens conjecture. *Journal für die reine und angewandte Mathematik* 357, 138–160.

Pogonowski, J. 2006. Projekt logiki infinitarnej Ernsta Zermela. *Investigationes Linguisticae* 14, 18–49.

Pogonowski, J. 2008. Metoda rezolucji i tablice analityczne w *Symbolic Logic* Lewisa Carrolla. *Investigationes Linguisticae* 16, 195–218.

Pogonowski, J. 2019. *Extremal axioms. Logical, mathematical and cognitive aspects*. Wydawnictwo Nauk Społecznych i Humanistycznych, Poznań.

Pogonowski, J. 2020. *Myślenie matematyczne. Drobne eseje przedemerytalne*. Wydawnictwo Nauk Społecznych i Humanistycznych, Poznań.

Posamentier, A.S., Lehmann, I. 2013. *Magnificent mistakes in mathematics*. Prometheus Books, Amherst (New York).

Steen, L.A., Seebach, J.A., Jr. 1995. *Counterexamples in topology*. Dover Publications, Inc., New York.

Wise, G.L., Hall, E.B. 1993. *Counterexamples in probability and real analysis*. Oxford University Press, New York.

Chapter 3

Standard, exception, pathology

3.1 Mathematical savoir vivre

Contrary to what a reader might expect from the title of this section, we are not going to discuss the behavior of mathematicians themselves. Instead our attention focuses on the following phrase which is very common in mathematical texts: *An object X is well behaving*. Here X may stand for a function, a topological space, an algebraic structure, and so on.

It needs to be clearly stressed that *well behaving* of mathematical objects is always related to some investigated theory or its applications. There is nothing like *absolute well behavior* – properties of objects are evaluated from a pragmatic point of view. As such we encounter declarations in the following vein specific domains of mathematics: Hausdorff spaces behave better than general topological spaces; Borel sets behave better than arbitrary sets; differentiable functions behave better than continuous functions; analytic functions (functions from the class C^ω) behave better than smooth functions (functions from the class C^∞); fields behave better than rings; recursive functions behave better than arbitrary functions; and Platonic solids behave better than arbitrary polyhedra, and so on.

It should also be mentioned that *well behaving* has nothing to do with *to be in majority*. Analytic functions are rare among continuous functions, almost every function is not differentiable anywhere, and there are only $\aleph_0$ recursive functions, etc. Platonic solids are very special convex polyhedra and are in minority, too. Thus, *well behaving* corresponds rather to a prototypic object, useful in a common practice. Such objects occurred as something like patterns at the beginning of the corresponding theory.

The concept in question is also related to the history of mathematics. Objects may become well behaving only when a theory underlying them is sufficiently developed. For example, complex numbers were viewed as well behaving only after proving that they form a field and providing a geometrical representation for them. Similarly for the negative numbers (which, together with natural numbers, form a structure closed under certain operations). Further examples include for instance fractals or purely random sequences.

It would be worth adding that there exists a certain limitation to the expressive power of the language of mathematics. Each theory uses at most countably many expressions because mathematical formulas are finite strings of symbols from at most countable vocabulary. On the other hand, the existence of uncountable sets is proved in many mathematical domains. Due to the mentioned limitation, only countably many of them can obtain *names* in the corresponding domain (meaning that they are denoted by closed terms from the language of the theory). One should strictly distinguish between *effective descriptions* and *definitions* of mathematical objects. For example, we can effectively describe only countably many ordinal numbers (using suitable systems of ordinal notation) while we can of course also define several uncountable ordinal (and cardinal) numbers.

The exposition of any mathematical theory usually starts with a description of objects which do behave well with respect to this theory or its applications. Then the presentation is extended in order to point out several *counterexamples*. There exist nice collections of counterexamples in main mathematical theories – see for instance Gelbaum and Olmsted 1990, 2003, Steen and Seebach 1995, and Wise and Hall 1993. The role of counterexamples is discussed in, for example, the classical work Lakatos 1976.

We distinguish *monomathematics* from *polymathematics* (the terms introduced in Tennant 2000). In monomathematics one deals with a specific mathematical structure – as for instance natural or real numbers. In polymathematics, in turn, one speaks about classes of structures – as for example groups, rings, fields, topological spaces, Hilbert spaces, Banach spaces, or graphs, and so on. It seems that it is in polymathematics where talking about well behaving objects makes any sense at all.

3.2 Standard

First of all, one should note that the concept of a *standard* object is not limited to the objects *called* standard in mathematical terminology. If an object *is* called standard by mathematicians, then this fact expresses their view that the object is somehow typical in the considered domain. There may exist, however, many other objects which deserve a qualification of being typical, prototypic, or normal, and so forth. In some cases the name *standard* is given to structures which are *intended* models of a given theory.

One of the important procedures in mathematics is to represent objects from a given domain in forms known as *normal*, *standard*, or *canonical*, among others. It is often the case that only after such a representation is one able to prove theorems about the initial – allegedly diversified – class of objects. Examples of the transformation in question are, among many others: conjunctive and disjunctive normal forms; prefix normal form; Skolem normal form (in logic); Cantor normal form (for ordinal numbers); Jordan normal form (for matrices over an algebraically closed field); canonical representation of integers (as products of powers of prime numbers); canonical differential forms (in analysis); and Kleene normal form (in recursion theory) and so on.

Another procedure, in a sense similar to the one mentioned above, is based on *representation theorems* – a very powerful tool in many mathematical domains. Examples of such theorems are:

1. *Stone Representation Theorem*. Each Boolean algebra is isomorphic to a field of sets.

2. *Cayley Representation Theorem*. Each finite group is isomorphic to some group of permutations.

3. *Mostowski Contraction Lemma*. Every extensional well-founded structure is isomorphic to a transitive structure.

4. *Nash Theorem*. Every Riemann manifold can be isometrically embedded in some Euclidean space.

5. *Whitney Representation Theorem*. For any m-dimensional differential manifold there exists its embedding into a $2m$-dimensional Euclidean space ($m > 0$).

6. *Gödel Representability Theorem*. Every recursive function is representable in Peano arithmetic.

In a sense, each completeness theorem in logic is a kind of a representation theorem. Algebraic characterization of elementary equivalence also serves as an example of representation of semantic notions.

Furthermore, several *classification procedures* may also be conceived of as bringing order to the domain in which we observe some regularities. Here are a few examples:

Classification of finite simple groups. Every finite simple group is isomorphic to one of the following groups: a cyclic group of prime order, an alternating group of degree at least 5, a simple group of Lie type (among which several further classifications are made), or one of the 26 sporadic simple groups.

Classification of simple Lie groups. Each simple Lie group either belongs to one of the infinite series A_n, B_n, C_n, D_n, or else is isomorphic with one of the five exceptional simple Lie groups, i.e. with one of: G_2, F_4, E_6, E_7, or E_8. The series correspond to: A_n series (special linear groups $SL(n+1)$), B_n series (special orthogonal groups) $SO(2n+1)$, C_n series (symplectic groups $Sp(2n)$), and D_n series (special orthogonal groups $SO(2n)$).

Classification of surfaces. Every closed compact connected surface without boundary is homeomorphic with one of the three following: two-dimensional sphere; connected sum of g thori ($g \geqslant 1$); or connected sum of k projective planes ($k \geqslant 1$).

Classification of Riemann surfaces. Each simply connected Riemann surface is conformally equivalent to one of the following surfaces: *elliptic* – the Riemann sphere $\mathbb{C} \cup \{\infty\}$; *parabolic* – the complex plane $\mathbb{C}$; or *hyperbolic* – the open disc $D = \{z \in \mathbf{C} : |z| < 1\}$.

Several *isomorphism theorems* also serve as a basis of classification in the sense that they describe certain structures in a unique way, for instance:

Frobenius Theorem. Every associative algebra over the field of real numbers is isomorphic with one of the fields $\mathbb{R}$ (real numbers), $\mathbb{C}$ (complex numbers) or $\mathbb{H}$ (quaternions).

Hurwitz Theorem. Any normed algebra with division is isomorphic either with $\mathbb{R}$, $\mathbb{C}$, $\mathbb{H}$ or $\mathbb{O}$ (octonions).

Ostrowski Theorem. Any field complete with respect to an Archimedean norm is isomorphic with either $\mathbb{R}$ or $\mathbb{C}$ and the norm is equivalent with the usual norm determined by the absolute value.

Pontriagin Theorem. Any connected locally compact topological field is isomorphic with either $\mathbb{R}$, $\mathbb{C}$ or $\mathbb{H}$.

Classifications are ubiquitous in mathematics. We form more and more abstract objects (such as quotient spaces and quotient algebras) starting from an initial universe and creating new objects as equivalence classes.

A very special standard structure in mathematics is the set of all natural numbers (with operations of addition and multiplication as well as with the successor operation and zero as a distinguished element). This structure precedes, both conceptually and historically, other number systems (rational numbers, integers, reals, and complex numbers). There are two ways in which one can create number systems:

Genetic. The concept of number is generalized, passing from one domain to another, wider domain. Operations which were not allowed in the starting domain (such as division of integers) become possible in a wider domain. Older domains become special cases of new domains. The domains are interconnected by homomorphisms, preserving the arithmetical operations. This was the way arithmetic was developed up to the middle of the 19th century.

Axiomatic. The modern approach to number systems is an axiomatic one: we specify the axioms which must be obeyed by (arithmetical) objects and operations on them. This approach originated with the works of Peano, Grassmann, Dedekind, Weber, Hilbert.

Before the "structural revolution" which took place in 19th century mathematics *non-standard* could mean *rare* from the point of mathematical practice or rejected on the basis of some intuitive beliefs. The Pythagoreans were shocked by non-commensurable magnitudes. Hence the irrationality of, say, $\sqrt{2}$ meant the presence of an "alien intruder" in the harmonious world of rational numbers. Similarly, early algebraists thought of $\sqrt{-1}$ as something belonging to mathematical fiction and not mathematical reality. This is reflected in the (unfortunate) name *imaginary number* introduced by Descartes and used also today. The common belief that Euclidean geometry describes accurately and adequately the physical world was responsible for the view that it is the only "true" geometry, and as such is the only standard geometry.

The discovery of non-Euclidean geometries, the development of abstract algebra (groups, rings, fields, and so on), the arithmetization of analysis – these great achievements of mathematics in the 19th century created new contexts in which the notion of *standard* might be used. It seems that recently by *standard* (objects or structures) one understands the objects or structures that are prototypic, most successful in applications, somehow central in the investigated domain. To a certain extent this understanding is reflected in the teaching of mathematics (at least below university level).

The standard model of Peano arithmetic is uniquely (up to isomorphism) characterized in a second-order language. Mathematicians themselves take for granted that it is a perfectly determined structure. The logicians, in turn, keep on recalling that the first-order Peano arithmetic is not complete (and not categorical in any infinite power). This theory has continuum distinct countable models. Moreover, it has 2^{κ} (the highest possible number of) different models in each infinite power κ. It is a *wild* theory. At the moment, non-standard models of arithmetic are not widely used in mathematics – with the notable exception of non-standard analysis, where they enable us to talk in a precise way of *infinitely small* (and *infinitely large*) numbers. Recall that these infinitesimals were accepted by Leibniz (though with only metaphysical and not mathematical explanation), then were almost eliminated from analysis in the time of Cauchy, Weierstrass and their contemporaries. It should be added, however, that infinitesimals were not completely forbidden at that time – they appeared in the investigations of non-Archimedean structures, for instance in the works of Hahn. It was only some sixty years ago when they obtained an interesting mathematical formulation (based on the notion of ultraproduct) proposed by Abraham Robinson.

The problem of establishing what is *standard* is more complicated in set theory (more exactly, in Zermelo-Fraenkel first order set theory, which – as it is claimed – may serve as a basis for all mathematics). The accepted linguistic usage requires that a model $\mathfrak{M}$ of this theory be called *standard*, if the element relation $\in^{\mathfrak{M}}$ of the model is the actual element relation $\in$ restricted to the universe of the model $\mathfrak{M}$. Each transitive model is standard and each standard model is well founded. Assuming the existence of a standard model implies the existence of a minimal standard model which is contained in all standard models.

However, set theory is far from being categorical. The method of forcing (Cohen 1966) can be used for the construction of many diversified models of set theory. "Normal" mathematicians (meaning those not working on set theory) seem to believe in the universe of true sets and do not worry about subtle incompleteness phenomena inseparable from set theory itself. It is not possible to predict the future of set theory. It may happen that it will be replaced by another (more fundamental?) theory. Or perhaps we will think of set theory in a way similar to that in which we currently think about such branches of mathematics as algebra or topology (see Mostowski 1967).

We draw a distinction between standard objects (as characterized above) and *generic* objects and generic properties. A property holding for "typical" objects from a given class is called *generic* and objects possessing it are called *generic* objects from this class (for instance: a generic matrix is invertible, a generic polynomial does not have a root at zero). In this sense, "generic" usually means the same thing as "almost all", where the latter term obtains a precise mathematical meaning, depending on the context. A dual concept to "almost all" is "negligible", the meaning of which is derived from the meaning associated with "almost all".

"Almost all" in measure theory (and also in probability theory) means "on a set with measure one". Here, a dual notion (negligible) is of course a set with measure zero. In topology "almost all" usually means "on an open dense set". A dual notion is a nowhere dense set. Furthermore, a meagre set (a set of the first category) is a countable union of nowhere dense sets. A comeagre set is one whose complement is meagre, which means that it is the intersection of countably many sets with dense interiors. Sets which are not meagre are called sets of the second category. It should be added that sets of second category need not to be comeagre – a set may be neither meagre nor comeagre (and then it is of second category). Observe that "almost all" in the sense of measure theory does not coincide with "almost all" in the topological sense: for example, the rational numbers form a set of Lebesgue measure zero and at the same time a dense subset in the set of all real numbers (with the usual topology).

In several domains of mathematics we talk about *degenerate* objects and *limiting* cases. An object is degenerate if some of its properties take the most possible values. This is not a precise definition, but it can be supplied with a lot of intuitive examples: a point is a degenerate circle (with radius equal to zero); a parabola can degenerate into two distinct or coinciding parallel lines; an ellipse can degenerate into a single point or a line segment; a hyperbola can degenerate into two intersecting lines; in inversive geometry a line is a degenerate circle (with infinite radius); a degenerate triangle is one with collinear vertices and zero area; a sphere with radius zero is a point; a degenerate sphere; a degenerate continuum consists of one point, examples of degenerate polygons are a digon and a monogon; in graph theory; a k-degenerate graph is an undirected graph in which every subgraph has a vertex of degree at most k; and a random variable which takes only one value has a degenerate distribution (in the case when this value is zero, we obtain Dirac delta function as the probability distribution).

In general, degenerate objects in some class belong to a class of simpler objects (for example a point as a degenerate circle). We distinguish between degenerate objects and objects called *exceptions*, to which we shall now turn.

3.3 Exception

What is an *exceptional* object in mathematics? First, we are not talking about exceptional objects in any *absolute* sense. Exceptions are always connected with some results established within a certain mathematical theory. The most typical situation occurs when one classifies objects (according to some chosen criteria) and a few objects do not fit into any of the classes, for instance:

Sporadic finite simple groups. The classification of all finite simple groups gives, besides a few infinite regular families of such groups, also 26 groups not showing any predictable pattern. You may call them exceptional groups, but from a purely formal point of view there is nothing strange in the fact that some classes of the classification contain only one member.

Special Lie groups. The five groups G_2, F_4, E_6, E_7, and E_8 are distinguished among all simple Lie groups, because they do not fit into any of the four patterns of all the remaining such groups. The properties of the groups from the four infinite series are well known. Many properties of the special groups are also known – in particular, there are numerous connections between these groups and concepts from theoretical physics, as has been demonstrated in recent decades.

The term *exceptional* can be applied not only to objects that lack properties taken into account during classifications, but also to such objects which are distinguished due to the possession of regularities absent in all other objects from a given class, for example:

Platonic solids in n dimensions. In three dimensions there are exactly five Platonic solids, as has been well known since antiquity. They are convex polyhedra with faces formed by regular polygons such that at each vertex the same number of edges meet. The Platonic solids are exceptional among convex polyhedra in the sense that they possess a lot of nice symmetries. Such regularly symmetric cells exist also in higher dimensions – for example in the fourth dimensions there are six of them, and in all higher dimension there are exactly three of them.

Standard model of Peano arithmetic. On the one hand, natural numbers are the most standard objects in mathematics. On the other, the standard model $\mathfrak{N}_0$ of Peano arithmetic PA is a very exceptional structure in the realm of all countable models of this theory; $\mathfrak{N}_0$ is the only well-ordered model of PA, $\mathfrak{N}_0$ is the only recursive model of PA (Tennenbaum theorem); and $\mathfrak{N}_0$ is a prime model of PA.

Other contexts in which one talks about exceptional objects include looking for *counterexamples*. For instance, the *Alexander horned sphere* is homeomorphic with the sphere S^2. However, it divides the space $\mathbb{R}^3$ in a different way to S^2: its inside is homeomorphic with the inside of S^2 but its outside is not homeomorphic with the outside of S^2. Hence it provides a counterexample to the Jordan-Schönflies theorem in three dimensions.

Still another situation when mathematicians talk about exceptions (or even pathology) is the absence of a recognizable pattern in a class (or series) of objects for which they expected to find such a pattern. However, not always does the lack of a pattern indicates pathology. The prime numbers form a very distinguished sequence of natural numbers. The concept of a prime number is very elementary – it is, as a matter of fact, recursive. There exist algorithms which establish whether a given natural number is prime or not. However, the distribution of primes remains one of the most mysterious problems of not only number theory but also some other branches of mathematics. One can prove that there are infinitely many primes of a given form, one can prove that there occur arbitrarily long arithmetic sequences consisting entirely of primes, and one can estimate the number of primes below a given natural number, and so on. However, one cannot say that we fully recognize the pattern according to which primes are allegedly distributed among all natural numbers.

3.4 Pathology

The very term *pathological* immediately implies some negative associations. Mathematical objects named *pathological* (sometimes also: *paradoxical*) appear as unexpected and, moreover, unwilling. There seem to be at least two typical situations in which one speaks about pathological objects in mathematics:

A clash with established intuitions. In a given epoch, mathematicians share intuitive views about the concepts they are dealing with. Discoveries of new kinds of objects may contradict these intuitions. But if the new objects appear fruitful in applications, if they are equipped with a sound theory, then the initial intuitions are forced to change.

1. *Infinity.* During centuries of evolution, mathematics continuously outpaced the fear of infinity. Actual infinity was treated with suspicion in ancient times, then it became more and more domesticated, so to speak. We can see this process in the development of calculus, in the work with infinite series, and in reaching the formally correct definition of a limit, and so on. Some properties of infinite collections seemed paradoxical (for example contradicting Euclid's postulate that a part is smaller than the whole), but then a change of perspective took place: the numerical equivalence of a set with its proper subset was taken as a formal definition of an infinite set.

2. *Numbers.* Mathematicians gradually overcame the resistance against new kinds of numbers which naturally arose in arithmetical and algebraic problems. Thus irrational, negative and imaginary numbers entered the realm of mathematics as fully-accepted objects only after they had been shown to behave properly – as structures closed with respect to prescribed operations.

New definitions of concepts formerly understood in an intuitive way. This situation arose for example in the case of a general definition of *function*. Formerly, by a function one understood a recipe according to which one value could be obtained as depending on other values. After accepting the most general definition – a function as a set of ordered pairs – plenty of *monsters* entered the stage. The first examples of continuous though nowhere differentiable functions were given by Bolzano and Weierstrass. "Space filling" functions were defined by Peano and Hilbert. All these monstra contradicted some intuitive beliefs regarding how (the diagram of) a function should look. This was the price for replacing intuitive views (based for example on concepts from outside mathematics, such as motion) with precise formal definition. This was, without a doubt, a justified solution: the development of mathematics is impossible without intuitions, but they should nevertheless always be abandoned when the deductive work begins.

Let us recall a much-quoted passage from Henri Poincaré's *Science and method* (Poincaré 2017, 132):

> Logic sometimes makes monsters. For half a century we have seen a mass of bizarre functions which appear to be forced to resemble as little as possible honest functions which serve some purpose. More of continuity, or less of continuity, more derivatives, and so forth. Indeed, from the point of view of logic, these strange functions are the most general; on the other hand those which one meets without searching for them, and which follow simple laws appear as a particular case which does not amount to more than a small corner.
>
> In former times when one invented a new function it was for a practical purpose; today one invents them purposely to show up defects in the reasoning of our fathers and one will deduce from them only that. If logic were the sole guide of the teacher, it would be necessary to begin with the most general functions, that is to say with the most bizarre. It is the beginner that would have to be set grappling with this teratologic museum.

Compare this quotation with the opinion expressed eighty years later by Freeman Dyson (Dyson 1978, 677–678):

> Modern mathematics began with Cantor's set theory and Peano's space filling curve. Historically, the revolution was forced by the discovery of mathematical structures that did not fit the patterns of Euclid and Newton.
>
> These new structures were regarded . . . as 'pathological', . . . as a 'gallery of monsters,' akin to the cubist painting and atonal music that were upsetting the established standards of taste in the arts at about the same time.
>
> The mathematicians who created the monsters regarded them as important in showing that the world of pure mathematics contains a richness of possibilities going far beyond the simple structures that they saw in nature. Twentieth century mathematics flowered in the belief

> that it had transcended completely the limitations imposed by its natural origins.
>
> Now, as Mandelbrot points out ... Nature has played a joke on the mathematicians. The 19th century mathematicians may have been lacking in imagination but Nature was not.
>
> The same pathological structures that the mathematicians invented to break loose from 19th century naturalism turns out to be inherent in familiar objects all around us.

Many objects considered as pathological or strange are already quite well described – a very nice example is the book Kharazishvili 2006 devoted to strange functions in real analysis.

A prototypic example of an object originally thought of as a pathological one and then becoming normal, standard, and "domesticated" is the Cantor set. In recent times, no professional mathematician has considered it a pathology. This is due to its fundamental role for instance in topology. Only in popular books do authors try to frighten innocent readers with devilish mysteries of the Cantor set and other fractal sets.

Traditionally, when speaking of paradoxical objects and constructions one recalls the Banach-Tarski theorem which – as is sometimes claimed – strongly violates our intuitions concerning volume and, in general, measure. It should be kept in mind, however, that the proof of this theorem is highly ineffective (axiom of choice is used) and the "pieces" into which the original ball is "cut" are non-measurable in the sense of Lebesgue. Hence, if one wants to talk about violation of intuitions in this case, then one should refer rather to the proof, and not to the theorem itself.

It is ridiculous – in our opinion at least – to expect that our common intuitions connected with everyday experience should be respected at any stage of sophisticated mathematical constructions. At any rate, there are cases in which operating with well-known objects, using perfectly natural proof techniques, one arrives at results which are strongly divergent from common intuitions:

Smale's theorem. This theorem states that it is possible to turn a sphere S^2 inside out in a three-dimensional space $\mathbb{R}^3$ (with possible self-intersections) without creating any crease. One speaks also of *eversion* of the sphere in this context. The proof of the theorem itself does not involve very sophisticated tools: one simply shows that there exists a regular homotopy of two immersions of S^2 into $\mathbb{R}^3$. Moreover, some physical models illustrating the eversion in question have been elaborated – one can find several videos online showing how the eversion is obtained.

Exotic spheres. We recall that an exotic sphere is a differentiable manifold M that is homeomorphic but not diffeomorphic to the standard Euclidean

n-sphere. This means that M is topologically indistinguishable from the Euclidean sphere, but admits a smooth structure which is essentially different from the standard such structure. The first exotic spheres were discovered by John Milnor: he proved that there exist distinct smooth structures on S^7 (there are 28 such structures on S^7). In some dimensions there are no exotic spheres (for example in dimensions: 1, 2, 3, 5, 6, and 12), while in others there exist thousands of them. Should we call exotic spheres pathological? If so, then we would be claiming to have a deep understanding of differential structures in very high dimensions, which does not seem to be the case today. The problem as to whether there exist exotic spheres in the fourth dimension remains open. If such an exotic 4-sphere were to exist, then it would be a counterexample to the smooth generalized Poincaré conjecture in dimension 4.

Exotic $\mathbb{R}^4$. We recall that an exotic $\mathbb{R}^4$ is a differentiable manifold that is homeomorphic to the Euclidean space $\mathbb{R}^4$, but not diffeomorphic to it. Let us also recall that dimension 4 is exceptional in this respect: exotic structures do not occur on $\mathbb{R}^n$ for $n \neq 4$. But $\mathbb{R}^4$ itself admits a continuum of distinct smooth structures on it. Again, we see that dimension 4 remains very mysterious.

We do hope that the above examples show that the concept of pathology in mathematics is pragmatically biased and relative to the development of mathematical theories. Let us also add two more comments:

Surprising results. Sometimes it is difficult to draw a boundary between unexpected surprises and pathologies in mathematics. For example, the existence of *pathological satisfaction classes* was, one may justly assume, a surprise. So on the one hand it is perhaps surprising that there are so many such classes and, on the other, that their properties contradict our intuitions concerning the relation of satisfiability.

Pathology versus mystery. One should remember that the non-existence of a recognizable pattern need not mean that one is surrounded by pathologies. It is not the case that everything we do not understand yet is automatically paradoxical or pathological. Mathematical knowledge does not form a closed system.

3.5 A final word

We admit that the above remarks bring no essentially new insights into the realm of mathematics. But we do hope they at least clarify some pragmatic components involved in the way we speak about mathematical objects: standard ones, exceptional, and pathological. The mathematical examples chosen in the paper are among the most typical, although they were in a sense selected randomly, without any claim to completeness. It might be an entertaining challenge to work on standards, exceptions and pathologies in separate branches of mathematics from a historical perspective.

Bibliography

Cohen, P. 1966. *Set theory and the continuum hypothesis*. New York.

Dyson, F. 1978. Characterising irregularity. *Science*, May 1978.

Gelbaum, B.R., Olmsted, J.M.H. 1990. *Theorems and counterexamples in mathematics*. Springer-Verlag, New York.

Gelbaum, B.R., Olmsted, J.M.H. 2003. *Counterexamples in analysis*. Dover Publications, Inc., Mineola, New York.

Kharazishvili, A.B. 2006. *Strange functions in real analysis*. Chapman & Hall/CRC, Taylor & Francis Group, Boca Raton, London, New York, Singapore.

Lakatos, I. 1976. *Proofs and refutations. The logic of mathematical discovery*. Cambridge.

Mostowski, A. 1967. Recent results in set theory. In: Lakatos, G., (Ed.), *Problems in the philosophy of mathematics, Proceedings of the international colloquium in the philosophy of science, London 1965*. North-Holland Publishing Company, Amsterdam, 82–108.

Poincaré, H. 2017. *Science and method*. Cosimo Classics.

Steen, L.A., Seebach, J.A., Jr. 1995. *Counterexamples in topology*. Dover Publications, Inc., New York.

Tennant, N. 2000. Deductive versus expressive power: a pre-Gödelian predicament. *Journal of Philosophy* 97, 257–277.

Wise, G.L., Hall, E.B. 1993. *Counterexamples in probability and real analysis*. Oxford University Press, New York.

Chapter 4

Intended and standard models

4.1 Introduction

One has recently been able to observe a revival in the discussion concerning intended models of mathematical theories. Its roots may be associated with the first axiomatizations of number systems and systems of geometry in the 19th century and at the beginning of the 20th century. Hardly any discussion of this kind is to be found earlier in the history of mathematics for obvious reasons: it was only in the 19th century that a systematic study of *several* structures began. Subsequently, one could then reasonably ask which of these structures can be uniquely characterized and which are the *intended* ones. Almost one hundred years ago mathematicians discovered *non-standard* models and later proved several *limitative theorems* connected with the possibilities of unique characterization of models. At about the same time certain *extremal axioms* were formulated in order to achieve such characterizations.

4.1.1 Innovations are not non-standard models

Obviously, the distinction between objects that are *normal* (*natural*, *standard*, etc.) objects and *unintended* (*unwilling*, *imaginary*, etc.) was also noticed earlier – just think for instance of negative numbers or imaginary numbers which were long rejected as legitimate mathematical objects before they finally became accepted by the mathematical community. Haim Gaifman discussed the following *innovations* in mathematics in his paper devoted to the non-standard models (Gaifman 2004): the discovery of irrationals; the incorporation of negative and complex numbers in the number system; the extension of the concept of *function* in the nineteenth century;

and the discovery of non-Euclidean geometry. Gaifman stresses that in the above cases we should not speak of non-standard models:

> Let me sum up the four historical cases and how they differ from non-standard models. First, the discovery of incommensurables is a discovery that a certain presupposition about spatial magnitudes was false. There is only one model, the standard one; we were simply mistaken about one of its basic features. Second, the enlargement of the positive number system by incorporating negative and complex numbers amounts to utilizing the possibilities inhering already in the positive numbers; there is no change of the standard model, but an unfolding of it. Third, the extension of the function concept to that of an arbitrary mapping (given as a set of pairs) is an explication of a previously loose concept, which is needed because new examples do not conform to previous expectations. There are no two models, but one developing conception. Fourth, the discovery of non-Euclidean geometry is the discovery that the concept of geometric space is ambiguous and admits an additional specification besides the received one; the difference is expressed as the denial of an accepted postulate. Here indeed there are several models and non-Euclidean geometry is, as Gauss noted, strange. If 'non-standard' is another word for 'strange' then it is "non-standard". But, as Gauss, Bolyai and Lobachevsky made clear, and as subsequent developments have borne out, non-Euclidean geometry is a legitimate conception of geometric space. If physical space is to be the arbiter of truth (as Gauss suggested) then neither the Euclidean nor the hyperbolic geometry is the winner. But whatever the verdict of physics, the different geometries constitute different specification of the general mathematical concept of geometric space. (Gaifman 2004, 13)

There are, of course, numerous further examples of such innovations in the history of mathematics. It may happen that new objects or structures are easily incorporated into the existing body of knowledge – such seems to have been the case with hypercomplex numbers introduced in the 19th century, with p-adic numbers, and with non-Archimedean structures. Or it may take a considerably longer time for the acceptance of new objects and structures (as in the cases of negative and complex numbers). In some cases we probably should modestly admit that we do not yet have a theoretical background developed sufficiently to see such new objects in a proper perspective (could this be the case with exotic spheres?).

Gaifman argues that some structures can be justly named *standard*. This is obviously the case with natural numbers. His further examples include *well orderings* and the class L of all *constructible sets*. On the other hand, the *full powerset operation* eludes the list of standard concepts. Even the full powerset of the set of all natural numbers seems to be inaccessible from the point of view of the linguistic and logical tools at our disposal.

4.1.2 The distinction: intended model – standard model

The terms *intended model* and *standard model* are sometimes used interchangeably in literature. However, we would like to propose the following distinction. By the *intended model* we will mean the structure which has been investigated for its own sake, typically for a long time, so that we have collected a large amount of data about it and have proven the most fundamental theorems about it. One could say that intended models are such structures which became *domesticated*, easily accessible cognitively and responsible for the basic mathematical intuitions. This is evidently only an intuitive characterization of the term *intended model* which, in turn, also is an intuitive term. Natural number series (with arithmetical operations) may serve as an example of an intended model understood in this way.

One can talk about *standard models* only after having a developed formal theory, ultimately an axiomatic one. In such a case the class of all possible models of the theory in question can be established. Then the standard model of such a theory is its model most closely related to the intended one. It may happen that standard models of a theory obtain precise characterization, for example in terms of isomorphism or elementary equivalence. Examples of standard models in the proposed sense are: natural number series (with arithmetical operations); the real field $\mathbb{R}$; and the model of Euclidean geometry (in Hilbert's axiomatization).

Is the geometric continuum standard? Well, this question depends on what we would like to consider a formal representation of geometric continuum. Feferman lists a few candidates (Feferman 2008): Euclid's continuum; Cantor's continuum; Dedekind's continuum; Hilbert's continuum; the continuum as the set of all branches in the full binary tree; and the continuum as the family $\wp(\mathbb{N})$ (the full powerset of the set of all natural numbers). Feferman summarizes his paper on conceptions of the continuum as follows:

> Of all the conceptions of the continuum considered here, only those of sec. 3 stand as structural ones, and of those only 2^N and $S(N)$ stand as *basic* structural conceptions. For, the continuum in Euclidean and Hilbertian geometry is not an isolated notion, while the continuum as given by Cantor's and Dedekind's construction of the real numbers, are hybrid constructions. The set 2^N of all sequences of 0s and 1s isolates the set-theoretical component of Cantor's construction, while the set $S(N)$ of all subsets of N isolates that of Dedekind's construction, but both of these lose entirely the basic geometric intuition of the continuum. On the other hand, it does not count against Cantor's

> and Dedekind's conceptions of the continuum in the form of the real number system R that they are hybrids of geometrical, arithmetical and set-theoretic notions. On the contrary, by a kind of miracle of synergy, R has proved to serve together with the natural numbers N as one of the two core structures of mathematics; together they are the *sine qua non* of our subject, both pure and applied.

According to the (intuitive, no doubt) characterization of the notion of a standard model proposed above, one can notice that:

1. Natural numbers do have the standard model: it is the only (up to isomorphism) Peano algebra.
2. Recently, the geometric (Euclidean/Hilbertian) continuum has been commonly represented with the use of the real numbers. Hence, $\mathbb{R}^3$ is thought of as the standard model of 3-dimensional geometry. But the debate on continuity is not over yet – and one can assume that some rival representation might replace the present one.
3. The ordered field $\mathbb{R}$ is unique, up to isomorphism. It is also the maximal Archimedean field. It is thus an example of a standard model. Similar arguments work for the complex field $\mathbb{C}$.
4. The situation in set theory is quite different, mainly due to the famous incompleteness results. Minimal axioms (axioms of restriction) in set theory have been rejected and known maximal axioms do not imply the existence of *the* standard model of set theory.

Suppose that we have chosen a model M of the theory T to be its standard model. Before T was formalized it had an intended model (of course this does not apply for group theory or theory of topological spaces, but it applies well to arithmetic of natural and real numbers). We dare to think about the connection between the intended model and the standard model of T as an analogy to the Church-Turing Thesis in computation theory. On the one hand we have an intuitive concept – namely the *intended model*, while on the other we have a well-defined formal concept – the *standard* model. But still, calling a model *standard* is a matter of choice depending on pragmatic criteria.

4.1.3 The existence of a unique characterization

There are, *a priori*, several possibilities for a unique characterization of a theory's interpretation:

Syntactic. This corresponds to the case when purely syntactic relations should establish exactly one interpretation.

Semantic. If you can show that all models are indistinguishable with respect to the sentences true in them, then the theory is *complete* – it has exactly one model up to the relation of elementary equivalence of models.

Algebraic. If you can show that all models are indistinguishable with respect to their structure, meaning that they are isomorphic, then the theory in question is *categorical.* On the one hand, mathematicians do not distinguish isomorphic models. Thus, if a (consistent) theory is categorical, then it means that it has exactly one model. But categoricity results require a rather strong logic, such as second-order. On the other hand, the condition in question can be weakened to *categoricity in power*. A (consistent) theory T is categorical in power κ if and only if it has a model in this power and all its models of power κ are isomorphic. This is a nice property and much can be said about the structure of models of such a theory.

Pragmatic. One can also go beyond the domain of deduction itself and impose some conditions with pragmatic content on interpretations of the theory. For example, one can demand that the interpretation should be as large as possible.

All these possibilities have been used in the characterization of models. Let us note the obvious fact that there are two kinds of mathematical theories: those which deal with an *a priori* given structure (arithmetic of natural numbers, integers, rational numbers, real real numbers, Euclidean geometry) and those which concern classes of structures (algebra, topology). In the first case we can reasonably talk about intended models, in the second it is more appropriate to talk about *prototypes*. Observe that, for instance, at the beginnings of topology mathematicians considered Hausdorff spaces the most general topological spaces: it was only later that the separation condition in question applied as determining a certain class of general topological spaces. The prototypes for general topological spaces would seem to have originated from the investigation of subsets of reals.

Hilbert noticed that the development of mathematics was based on two methods: the *genetic method* and the *axiomatic method*. The former was typical for the extension of the concept of number, while the latter was applied before Hilbert's time, initially (if not only) to geometry. The expansion of the axiomatic method began in the second half of the nineteenth century.

In the case of polymathematics (dealing with classes of models and not with some structure given in advance) we should not talk about intended models. However, the characterization of models can be obtained here by several classification and representation theorems as well as by looking for lists of invariants.

It is customary in mathematical discourse to talk about *well behaving* objects – in contradistinction to objects which behave *badly*, which are *exceptions* or *pathologies*. We have written about that in the essay *Standard, exception, pathology* included in this volume.

The origin of two metalogical notions, *categoricity* and *completeness*, is related to works of mathematicians and logicians from the 19th and 20th centuries (Awodey and Reck 2002a, 2002b, Corcoran 1980, 1981). The most important steps in this development are the following:

1. Axiomatizations of natural numbers by Peano and Dedekind.
2. Hilbert's axiomatization of geometry.
3. Huntington's and Veblen's proposals concerning categoricity and completeness.
4. Fraenkel's reflections on categoricity and completeness in set theory.
5. Incompleteness results by Gödel and others.
6. Characterizations of categoricity and completeness in classical and modern model theory.

Limitative theorems proven in the twentieth century revealed objective restrictions concerning the unique characterization of models of relatively rich theories. This topic is of course well known and there is no need to describe it in detail here. Let us only mention that among the most important results in this respect are: the Löwenheim-Skolem theorem; Skolem's construction of a non-standard model of arithmetic; Gödel's incompleteness theorem; and independence results in set theory. We have discussed these issues in more detail in the book Pogonowski 2019, which we draw from in the material that follows.

4.2 Extremal axioms

The term *extremal axiom* (to our knowledge first used in Carnap and Bachmann 1936) has its origins in Hilbert's *Grundlagen der Geometrie*. Hilbert added to his system of axioms for geometry an extra axiom, the *axiom of completeness*, which was supposed to express the idea that the underlying system of geometrical objects (points, lines and planes) is unable to have any extension without violation of the remaining axioms. In this form, the axiom of completeness was not a statement of the object language of the theory; it belonged to metalanguage, as expressing some view about models of geometry. It has been replaced by the *axiom of continuity* in the later versions of *Grundlagen der Geometrie* (Hilbert 1988).

Carnap and Bachmann tried to obtain a general scheme of extremal axioms. They write at the beginning of their paper:

> Some important axiom systems are so constructed that first a series of axioms is given, making certain statements about the basic concepts of the axiomatic theory, and then at the end an axiom of a special sort appears which apparently speaks about the foregoing axioms and not about the special concepts of the theory. The most famous axiom system of this sort is Hilbert's axiom system of Euclidean Geometry. It ends with the famous "completeness axiom" which runs as follows [The footnote given here by the authors reads: D. Hilbert, *Grundlagen der Geometrie* (Leipzig and Berlin). We take the Hilbert completeness axiom in the form it has in editions 2–6, not the 'linear formulation' of the 7th edition of 1930.]:
>
> 'The elements (points, lines, planes) of geometry constitute a system of things which cannot be extended while maintaining simultaneously the cited axioms, i.e., it is not possible to add to this system of points, lines, and planes another system of things such that the system arising from this addition satisfies axioms AI-V1.'
>
> Axioms of this sort, which ascribe to the objects of an axiomatic theory a maximal property – in that they assert that there is no more comprehensive system of things that satisfies a given series of axioms – we call a maximal axiom. The same axiomatic role as that of maximal axiom is played in other axiom systems by minimal axioms which ascribe a minimality property to the objects of the discipline. Maximal and minimal axioms we call collectively extremal axioms. (Carnap and Bachmann 1981, 68–69)

Carnap and Bachmann worked in the simple theory of types. They defined several auxiliary notions, such as complete isomorphism, dividable and undividable structures, and so on. They formulated four types of extremal axioms: minimal and maximal, and applicable either to models or structures (where structure was understood as equivalence class of models with respect to complete isomorphism). The axioms discussed in the paper included Hilbert's completeness axiom in geometry, and Fraenkel's restriction axiom in set theory and the axiom of induction in arithmetic. They discussed also the problems of categoricity and completeness of theories. The project as a whole was very ambitious, although it would seem not to have been very influential at that time. Some of the issues discussed by Carnap and Bachmann were also investigated for instance in Lindenbaum and Tarski 1936 and Tarski 1940. There exist only relatively few works devoted to the extremal axioms in general. Those worth mentioning include: Awodey and

Reck 2002a, 2002b, Hintikka 1986, 1991, Schiemer 2010a, 2010b, 2012, 2013, Schiemer and Reck 2013, Schiemer, Zach and Reck 2015, and Tarski 1940. In the book Pogonowski 2019 we have taken into consideration several further extremal axioms: the axiom of continuity in algebra; restriction axioms in set theory (besides Fraenkel's axiom also Gödel's axiom of constructibility, von Neumann's axiom of the limitation of size and Suszko's axiom of canonicity); and maximal axioms in set theory, that is axioms of the existence of large cardinal numbers and axioms generalizing Dedekind's idea of cuts.

4.3 Natural numbers

The natural numbers (together with arithmetical operations on them) form the most fundamental structure in mathematics. They originated in the process of abstraction from reflections about magnitudes. Thus the intended model in this case is the most salient example of what can be called an intended model. The data about numbers have been collected for millennia. Our views on about numbers were shaped by the theory of proportions contained in Euclid's *Elements* practically up to the beginning of the nineteenth century.

4.3.1 Looking for a unique characterization

Categorical second-order characterizations of natural numbers were proposed in the second half of the nineteenth century by Peano and Dedekind. They both made use of the concept of set, which at that time was quite new. It should probably be added that even before them an axiomatic description of natural numbers had been proposed by Hermann Grassmann.

Peano used the axiom of induction formulated in the language of second-order logic. Mathematical induction is a fundamental pattern of inference. Simultaneously, the induction axiom is thought of as a restriction on models of the theory, with the aim of eliminating *alien intruders*.

Mathematicians are satisfied with the fact that there exists exactly one (up to isomorphism) Peano algebra $(N, s, 0)$ (with the universe N, the successor function s and the initial object 0) and, moreover, that for any two Peano algebras there exists exactly one isomorphism between them.

Dedekind, in his *theory of chains*, also obtained a unique characterization of natural numbers. It was observed (Hao Wang, analyzing Dedekind's letter to Keferstein) that Dedekind was aware of the necessity of the elimination of *alien intruders* from the universe of natural numbers – see a more detailed discussion in Di Giorgio 2010.

The first non-standard model of arithmetic was constructed by Thoralf Skolem in 1933-1934. It is an interesting application of the prototype of an *ultraproduct* construction. It could be interesting to investigate more closely Skolem's inspirations. On the one hand, he produced a non-standard model of set theory in 1922, as a consequence of the Löwenheim-Skolem theorem. On the other, his construction of a non-standard model of arithmetic could be compared with the *degrees of infinity* investigated earlier by Du Bois-Reymond and Hardy.

There are still other methods of obtaining non-standard models of arithmetic: one can use the compactness theorem or refer to the full binary tree of extensions of first-order Peano arithmetic PA, obtained by the addition of undecidable sentences to PA, in the well-known manner. Incompleteness results about arithmetic imply that there are continuum non-elementarily equivalent models of Peano arithmetic and hence also a continuum of pairwise non-isomorphic such models.

Besides a unique characterization of PA in second-order logic, one can also obtain such characterizations using some logical tricks: one can, for instance, expand the language by adding generalized quantifiers to it or use some infinitary logic.

4.3.2 Tennenbaum's theorem

The following theorem is sometimes used to support the claim that the intended model of arithmetic is nothing other than its standard model:
TENNENBAUM'S THEOREM. *Let $\mathfrak{M} \models PA$ be nonstandard. Then $\mathfrak{M}$ is not recursive.*

The paper by Quinon and Zdanowski uses this kind of argumentation (Quinon and Zdanowski 2007). By a *standard* model (of arithmetic) they mean a model with ordering of the type ω. On the other hand, a model that reflects our intuitions about numbers adequately they call an *intended* model. The authors make some assumptions which they call *cognitive*: the most basic feature of natural numbers is that we can count with them, and any property that humans can compute can also be computed by Turing machines. The second assumption gives an upper bound on what we humans may compute. Furthermore, they claim that we can identify only recursively isomorphic models and not arbitrary isomorphic models. They formulate the following two postulates:

1. The intended model of arithmetic has to be recursive.

2. The intended model of arithmetic has to satisfy induction (at least first-order induction).

They claim that induction together with Tennenbaum's theorem describes exactly the class of intended models of arithmetic. One should also mention that some writers do not accept argumentation such as that presented above. Identification of the intended model with the standard model *via* the Tennenbaum's theorem faces the problem of definition of recursivity (in either of many equivalent forms, for instance Turing machines or primitive recursive functions). The very idea of *finititude* is inseparably connected with all representations of recursivity. For a more detailed discussion, see for example Button and Smith 2012.

4.4 Continuity

Compared to the case above (the discrete universe of natural numbers), the case with an intended model of the *continuum* is more complicated. Early discussion of this concept was limited to the *geometric continuum*. Only later was continuity in an algebraic sense, meaning connected with number systems, taken into account.

4.4.1 Continuity in geometry

The history of the debate about the geometric continuum reaches over two thousand years in the past. For its classical exposition see for instance Boyer 1949. Quite recently, Baldwin proposed that the *geometric continuum* be understood as the line situated in the context of the plane (Baldwin 2015). This paper (divided into two parts, in the final version) discusses the changes in the axiomatization of the geometric continuum with special emphasis put on the works by Euclid, Descartes, Hilbert and Tarski. Let us briefly summarize some of its main points. According to Baldwin, Hilbert faced several challenges in his system of geometry:

1. To identify and fill 'gaps' or remove 'extraneous hypotheses' in Euclid's reasoning.

2. To reformulate propositions such as *Triangles and parallelograms which are under the same height are to one another as their bases* to reflect 19th century understanding of real numbers as measuring both length and area.

3. To ground the geometry of Descartes.

4. Additionally, Hilbert accomplished another task: grounding calculus.

Descartes' geometry made it possible to extract n-th roots for all n and provided solutions to many higher degree polynomials. In a sense, Cartesian geometry may be identified with real algebraic geometry: the study of polynomial equalities and inequalities in the theory of real closed fields. One should nevertheless remember that Cartesian constructions concerned *segments*, and lines are not conceived of as *sets* of points.

The language of Tarski's system of elementary geometry, also discussed by Baldwin, is a one-sorted language of relations on points. Tarski formulates the *schema of continuity*, which can be understood as a definable version of Dedekind cuts:

$$\exists a \forall x \forall y[\alpha(x) \wedge \beta(y) \rightarrow B(axy)] \rightarrow \exists b \forall x \forall y[\alpha(x) \wedge \beta(y) \rightarrow B(xby)],$$

where α and β are first-order formulas, α does not contain any free occurrence of a, b, y and β does not contain any free occurrence of a, b, x and B is a ternary relation of *lying between*.

Baldwin notices that Archimedes' axiom serves several purposes in Hilbert's *Grundlagen der Geometrie* (the first is essential, the other two – according to Baldwin – could be replaced by other ideas with more intuitive geometrical content):

1. Metamathematical: concerning the consistency and independence of the axioms.

2. It is used to justify the coordinatization of n-space by n-tuples of real numbers.

3. It is necessary in order to show that equicomplementable (equal content) is the same as equidecomposable (in 2 or more dimensions).

Baldwin puts forward several arguments in favor of the thesis that Dedekind's axiom (in Hilbert's system in the form of his *Vollständigkeitsaxiom*) is inappropriate as an attempt to axiomatize the Euclidean or Cartesian or even Archimedean geometric data, for instance the following:

1. The requirement that for any circular arc there exists a straight-line segment that could measure this arc contradicts the intentions of Descartes.

2. The continuity axiom does not belong to the geometric data – it is rather an external limiting principle.

3. Euclid implicitly assumed that there is only one geometry. But this very statement is not geometrical itself. The original formulation of

> the continuity axiom in *Grundlagen der Geometrie* was not of a purely geometrical nature – the axiom should imply that the model of all the remaining axioms is maximal.

Baldwin notices that among Hilbert's goals when proposing the axiom of continuity was that of providing a secure foundation for calculus. Finally, Baldwin compares his approach with that of Ehrlich:

> In a sense, our development is the opposite of Ehrlich in [28] (i.e. Ehrlich 2012). Rather than trying to unify all numbers great and small, we are interested in the minimal collection of numbers that allow the development of geometry according with our fundamental intuitions. [...]
> Philip Ehrlich has made several important discoveries concerning the connections between the two 'continuity axioms' in Hilbert and develops the role of maximality. First, he observes [...] that Hilbert had already pointed out that his completeness axiom would be inconsistent if the maximality were only with respect to the first order axioms. Secondly, he [...] systematizes and investigates the philosophical significance of Hahn's notion of Archimedean completeness. Here the structure (ordered group or field) is not required to be Archimedean; the maximality condition requires that there is extension which fails to extend an Archimedean equivalence class. [Footnote: In an ordered group, a and b are *Archimedes-equivalent* if there are natural numbers m, n such that $m|a| > b$ and $n|b| > a$.]

We have chosen Baldwin's proposal as an example of recent reflections on the structure of the geometric continuum. Of course, there are numerous other reflections on the same topic.

4.4.2 Reals, hyperreals, surreals

The intended model of *arithmetic continuum* should be identified with the structure described (partly intuitively) by the knowledge on real numbers collected through the millennia. In the nineteenth century several attempts at the axiomatization of real numbers were made. This history is well known from the literature and there is no need to recall the details here. The most influential were proposals by Cantor and Dedekind (and later by Hilbert), but one should also remember about other attempts, for instance by Weierstrass, Heine, Méray, and Weber.

Dedekind's construction of real numbers with his famous cuts (of the ordered structure of rational numbers) became an ubiquitous pattern in many domains (*Dedekind's completion*). Dedekind wrote in his famous treatise (Dedekind 1872):

> I find the essence of continuity ... in the following principle: 'If all points of the straight line fall into two classes such that every point of the first class lies to the left of every point of the second class, then there exists one and only one point which produces this division of all points into two classes, this severing of the straight line into two portions'.

The arithmetic continuum, as presented by Cantor or Dedekind, became the fundamental structure in many domains of mathematics, notably in the foundations of analysis. It seems that when one asks today a "normal" mathematician (meaning one not working in the foundations of mathematics) what *the* structure of the continuum is, then they most likely will propose the completely ordered (and maximal Archimedean) field $\mathbb{R}$ as the standard model of this concept.

It is often claimed that the program of *arithmetization of analysis* (due essentially to Weierstrass and others) provided the ultimate answer to the question about the structure of the continuum. In addition, it is also claimed that this program has eliminated *infinitesimals* from mathematics. This is not entirely true because the mathematical theory of infinitesimals and non-Archimedean structures was also developed, for instance in the works of Hahn, Stolz and others. In the 20th century the non-standard analysis originated by Abraham Robinson became a serious alternative for the classical approach to analysis.

There exist also conceptions of the continuum based on specific philosophies of mathematics, such as Weyl's characterization in terms of predicative definability (Weyl 1918) or Brouwer's characterization in terms of his intuitionistic theory of free choice sequences (Brouwer 1927).

Quite recently, Ehrlich wrote several papers devoted to the significance of the non-Archimedean structures and to the universal extensions of arithmetic continua. Let us summarize briefly some of his findings:

1. First of all, Ehrlich recalls facts concerning the emergence of non-Archimedean systems of magnitudes. They include proposals by: Du Bois-Reymond, Hausdorff, Hahn, Stolz, and Veronese.

2. The existence of this rich world of systems, so intensively investigated in the second half of the nineteenth and at the beginning of the twentieth century, contradicts the misconception claiming that infinitesimals were banished from mathematics once and for all at that time.

3. Ehrlich discusses the notion of *Archimedean completeness*. Recall that: in an ordered group, a and b (different from 0) are *Archimedes-equivalent* if there are natural numbers m, n such that $m|a| > b$ and $n|b| > a$.

Now, let G and G' be two ordered abelian groups and $G \subseteq G'$. We say that G' is an Archimedean extension of G, if for each non-zero $y \in G'$ there is an $x \in G$ which is Archimedean equivalent to y. If G admits no proper Archimedean extension (in which case we say that G satisfies *Hahn's Completeness Condition*), then G is said to be *Archimedean complete*. Hahn has realized (Hahn 1907) that unlike Hilbert's completeness condition, his completeness condition does not presuppose that G is Archimedean. Thus it is possible that there are non-Archimedean, Archimedean complete ordered abelian groups (in fact there are plenty of them).

4. Ehrlich argues that Conway's surreal numbers No (*all numbers great and small*) may be regarded as an *absolute arithmetic continuum* (modulo the von Neumann-Bernays-Gödel set theory NBG). He proves several results which confirm the thesis that:

   (a) The real number system should be regarded as constituting *an Archimedean arithmetic continuum*. $\mathbb{R}$ is (up to isomorphism) the unique real-closed field having no gaps modulo the Archimedean condition.

   (b) The class of surreal numbers No is (up to isomorphism) the unique real-closed ordered field having no gaps that are definable in terms of sets of NBG.

5. He uses some tools from classical model theory (namely, homogeneous universal models, and saturated models) which, together the Archimedean axiom and the theory of real-closed fields that are η_α-orderings, enable him to obtain novel (categorical) axiomatizations of the ordered field of reals.

An original and at the same time precise representation of infinitesimals was proposed in the nineteen-sixties by Abraham Robinson. His *non-standard* analysis introduced *hyperreal numbers*. The construction uses the notion of ultraproduct. This can be viewed as the first precise representation of Leibniz's infinitesimals. For some historical details, see for example Bell 2005. The field in question is real closed but it is not Archimedean.

Another approach to continuity was initiated by Lawvere in his lectures in 1967 (Lawvere 1979). In this kind of infinitesimal analysis a "nonpunctiform" or "smooth" modification of the real number system is considered. The underlying logic is no longer classical – one is forced to use intuitionistic logic instead.

4.4.3 Isomorphism theorems

The dominating role of $\mathbb{R}$ and $\mathbb{C}$ in analysis is obvious to everyone: classical calculus and complex function theory are based on these structures. Hence they can justly be named intended models. $\mathbb{R}$ is the only (up to isomorphism) completely ordered field. It is also the maximal Archimedean field but it is not algebraically closed. On the other hand, the field $\mathbb{C}$ cannot be ordered but has the following salient features:

1. $\mathbb{C}$ is algebraically closed.
2. $\mathbb{C}$ has characteristic 0.
3. The transcendence degree of $\mathbb{C}$ over $\mathbb{Q}$ equals continuum.
4. One can show that the above three properties characterize the structure $\mathbb{C}$ uniquely (up to isomorphism).

There are several further purely mathematical arguments supporting the distinguished role played by $\mathbb{R}$ and $\mathbb{C}$ and some other structures, for instance:

1. *Frobenius Theorem*. Each associative algebra with division over $\mathbb{R}$ is isomorphic either with $\mathbb{R}$, $\mathbb{C}$, or $\mathbb{H}$.
2. *Hurwitz Theorem*. Any normed algebra with division is isomorphic either with $\mathbb{R}$, $\mathbb{C}$, $\mathbb{H}$ or $\mathbb{O}$.
3. *Ostrowski Theorem*. Any field complete with respect to an Archimedean norm is isomorphic with either $\mathbb{R}$ or $\mathbb{C}$ and the norm is equivalent with the usual norm determined by the absolute value.
4. *Pontriagin Theorem*. Any connected locally compact topological field is isomorphic with either $\mathbb{R}$, $\mathbb{C}$ or $\mathbb{H}$.
5. *1,2,4,8–Theorem (Bott, Milnor, Kervaire)*. Each algebra with division over $\mathbb{R}$ has dimension 1, 2, 4 or 8.
6. *Hopf Theorem*. Each commutative algebra with division over $\mathbb{R}$ has dimension $\leqslant 2$.

It may be worth adding that when we pass from the field $\mathbb{Q}$ of rational numbers and $\mathbb{R}$ of real numbers to more general fields ($\mathbb{C}$ of complex numbers, quaternions $\mathbb{H}$, octonions $\mathbb{O}$, sedenions $\mathbb{S}$, Cayley-Dickson contruction or Clifford algebras) then we loose certain properties of multiplication in these structures:

	$\mathbb{R}$	$\mathbb{C}$	$\mathbb{H}$	$\mathbb{O}$	$\mathbb{S}$	
dimension	1	2	4	8	16	> 16
ordering	yes	no	no	no	no	no
commutativity	yes	yes	no	no	no	no
associativity	yes	yes	yes	no	no	no
alternativity	yes	yes	yes	yes	no	no
power associativity	yes	yes	yes	yes	yes	yes
zero divisors	no	no	no	no	yes	yes

4.5 Sets

Set theory is a relatively new mathematical discipline as compared for instance with arithmetic, algebra or calculus. One speaks about *standard* models of this theory but this seems to be only a linguistic clarification. If set theory is consistent, then it has a lot of models, including countable models which – though extensively used for example in the forcing technique – can hardly be considered adequate representations of the whole set theoretical universe.

Two types of extremal axioms were considered in set theory: minimal axioms (expressing the idea that the universe of all sets should be as narrow as possible) and maximal axioms (expressing the idea that the universe of all sets should be as rich as possible).

4.5.1 Minimal axioms

Several versions of minimal axioms have been proposed:

Fraenkel's Axiom of Restriction (Axiom der Beschränkheit). This asserts that only sets whose existence can be proved from the other axioms actually exist (Fraenkel 1928). Fraenkel's aim in proposing this axiom is connected with the elimination of some *unwanted* sets – as for instance not well-founded sets. Fraenkel also considered the relation of this axiom to possible notions of completeness, including categoricity, semantic completeness and decidability. A very detailed analysis of Fraenkel's ideas is contained in Schiemer 2010a, 2010b. In particular, Schiemer argues that Fraenkel's formulation of his axiom was inspired by Dedekind's notion of simply infinite chains.

Von Neumann's Axiom of the Limitation of Size. The idea that some collections are too big to be sets was well known at the very beginning of set theory, and the *limitation of size principle* was a heuristic principle which recommended the exclusion of such collections from the universe of all sets. There remained the problem of a precise formulation of this principle. Von

Neumann proposed that any object in the domain of the theory which is "too big" to be a set be considered similar to the class of all sets, in the sense that there exists a bijection between it and the class of all sets (von Neumann 1925).

Gödel's Axiom of Constructibility. This axiom was introduced by Gödel in order to prove the consistency of the axiom of choice and the continuum hypothesis with the remaining axioms of Zermelo-Fraenkel set theory (Gödel 1940). The hierarchy of constructible sets in Gödel's sense is defined by transfinite induction, starting with the empty set. Successor stages of the construction see the forming of the family of all definable (with parameters) subsets of the previous stage, and the union of all previous stages is taken at limit stages. The axiom of constructibility then asserts that each set is constructible, that is it belongs to some stage of the constructible hierarchy. Constructible sets can be also characterized by closure conditions under certain operations. Gödel later changed his attitude and argued in favor of maximality axioms in set theory, with an explicit reference to Hilbert's axiom of completeness in geometry.

Suszko's Axiom of Canonicity. Suszko aimed for an explication of the Skolem's (alleged) Paradox without any reference to the Löwenheim-Skolem Theorem (Suszko 1951). The main idea here was to build metasystems in which every object (set) has a name; such systems were called *canonical*. Suszko's proposal is partly based on Gödel's ideas. A different approach to the same problem was proposed by Myhill (Myhill 1952).

In Fraenkel, Bar Hillel and Levy 1973 the authors discuss the general form of two axioms of restriction:

1. First Axiom of Restriction: *Given a property Q of sets closed under set-forming operations, one claims that every set has the property Q.*
2. Second Axiom of Restriction: *All sets are constructible (in Gödel's sense) and there is no transitive set-model of ZF.*

The First Axiom of Restriction is equivalent to the conjunction of the axiom of regularity (the axiom of foundation) and the sentence saying that there are no strongly inaccessible cardinals. If we consider set theory in a second-order language with a suitable version of the First Axiom of Restriction, then we can prove categoricity of such a theory. However, no consequences concerning the continuum hypothesis can be drawn from the First Axiom of Restriction.

The Second Axiom of Restriction implies the continuum hypothesis. It also implies that there are no strongly inaccessible cardinals. Therefore, it is stronger than the first axiom of restriction.

4.5.2 Critique of minimal axioms

The axioms of restriction have been criticized from the very beginning. The most elaborate critique of them is proposed in Fraenkel, Bar Hillel and Levy 1973. The authors also formulated several arguments in favor of the maximality axioms in set theory, as for instance these contained in the following passage:

> In the case of the axiom of induction in arithmetic and the axiom of completeness in geometry, we adopt these axioms not because they make the axiom systems categorical or because of some metamathematical properties of these axioms, but because, once these axioms are added, we obtain axiomatic systems which perfectly ***fit our intuitive ideas about arithmetic and geometry***. In analogy, we shall have to judge the axioms of restriction in set theory on the basis of how the set theory obtained after adding these axioms ***fits our intuitive ideas about sets***. To restrict our notion of set to the narrowest notion compatible with the axioms of ZFC just for the sake of economy is appropriate only if we have absolute faith that the axioms of ZFC (and the statements they imply) are the only mathematically interesting statements about sets. It is difficult to conceive of such absolute faith in the sufficiency of the axioms of ZFC (as one would have in, say, the full axiom of comprehension if it were not inconsistent). Even if one had such a faith in the axioms of ZFC, ***it is likely that he would settle rather for something like the axiom of completeness, if there were some reasonable way of formulating it***. (Fraenkel, Bar Hillel and Levy 1973, 117) Emphasis: JP.]

The authors provide further arguments against axioms of restriction, related to mathematical elegance and the Platonistic point of view.

4.5.3 Maximal axioms

In the commentaries to his second axiomatization of set theory, Zermelo expressed serious doubts about the adequacy of approaches in which one talks about single models of set theory, in particular about countable such models (Zermelo 1930). He argued that set theory should be represented by an hierarchy of his *normal domains* and postulated the existence of a transfinite hierarchy of strongly inaccessible cardinal numbers. His normal domains can be characterized uniquely (in a strong system of logic) by two parameters, corresponding to the number of urelements (forming the base of the domain) and a strongly inaccessible cardinal number (corresponding to the height of the domain). He proved the isomorphism theorems for such domains.

Andrzej Mostowski expressed critical views on strong infinity axioms in Mostowski 1967. He also classified the strong infinity axioms into two groups:

1. *Passing from potential to actual infinity*. The universe of sets is potentially infinite and closed under certain operations on sets, so we may propose the existence of a new set embracing this whole universe and closed under these operations. In this way we obtain, for example, strongly inaccessible cardinal numbers.

2. *Existence of residual sets*. Suppose that during the generating of sets with the use of set theoretical operations we always obtain sets having some property P. If there are no evident reasons supporting the claim that all sets have the property P, then we may consider a new axiom stating the existence of sets without the property P. In this way we obtain for example measurable cardinals.

One should note that there are conflicts between maximal and minimal axioms. For example, the existence of a measurable cardinal implies the negation of Gödel's axiom of constructibility. Mathematical arguments in favor of large cardinals come from different domains of mathematics, most notably from measure theory.

Axioms stating the existence of large cardinals throw some light on the possibilities of provability and consistency strength of theories. Axioms of existence of large cardinals do not resolve the generalized continuum hypothesis. Thus, they are investigated for their own sake, for their applications in mainstream mathematics and above all for a better understanding of such metalogical notions as the consistency strength of theories.

It is not reasonable to formulate any categorical claims about the future of set theory. About half a century ago Mostowski discussed the idea that perhaps in the future set theory would share the fate of general topology. Anyway, the existence of many unsolved problems in set theory clearly points to the fact that it might be replaced by a theory more closely determining its fundamental concepts so as to obtain analogons of categoricity or completeness in some new sense.

4.6 Model theory

The first result in model theory (obtained even before the term *model theory* came into existence) is the Löwenheim-Skolem theorem. This was a negative result concerning the categoricity of consistent theories. Some positive early results include Cantor's characterization of a countable dense ordering without endpoints (categorical in power $\aleph_0$).

Classical model theory, which originated with the works of Alfred Tarski, Robert Vaught and others, brought several results about complete theories, for example the Łoś-Vaught test: if T is a consistent theory without finite models which is κ-categorical in some infinite cardinal κ, then T is complete. This result was used, for instance, for the proof of the following theorems:

1. The theory of atomless Boolean algebras is $\aleph_0$-categorical.

2. The theory of algebraically closed fields of characteristic 0 (or p, where p is a prime number) is categorical in any uncountable power

3. The theory of infinite abelian groups in which each element has the rank p is κ-categorical for any κ.

There are numerous other theorems concerning complete theories and their models as well as other model-theoretic concepts, for example *model completeness*. The notion of $\aleph_0$-categoricity (meaning categoricity in power $\aleph_0$), for instance, is characterized by the following theorem which connects this notion with the space of n-types:

The Ryll-Nardzewski-Engeler-Svenonius theorem. The following conditions are equivalent for any (consistent) theory T:

1. T is $\aleph_0$-categorical.

2. For every natural number n, T has only finitely many n-types.

3. For every natural number n, every n-type is isolated.

4. For every natural number n, the Stone space $S_n(T)$ is finite.

In 1954 Łoś formulated the conjecture that the only possibilities of categoricity (in power) of theories are the following:

1. $(+,-)$: T is $\aleph_0$-categorical, but is not κ-categorical for all uncountable κ. Example: the theory of dense linear order without the first and last elements.

2. $(-,+)$: T is not $\aleph_0$-categorical, but is κ-categorical for all uncountable κ. Example: the theory of algebraically closed fields of characteristic 0 (or p, where p is a prime number).

3. $(+,+)$: T is κ-categorical for all cardinals κ. Example: the pure theory of identity.

4. $(-, -)$: For all cardinals κ, T is not κ-categorical. Example: the theory of the standard model of PA.

In 1965 Morley positively confirmed Łoś's conjecture. Morley's theorem (if a theory is categorical in *some* uncountable power, then it is categorical in *all* uncountable powers) was a starting point for modern model theory. The main emphasis here is put on spectra of theories, investigations of the space of types, and definability problems. Shelah obtained several results concerning the classification of theories with respect to the number of their non-isomorphic models. It was shown that theories can be classified into: ω-stable, superstable, strictly superstable or unstable.

As an example of investigations of categoricity let us recall Grzegorczyk's paper written half a century ago (Grzegorczyk 1962). The author discusses several meanings connected with the concept of *categoricity*, which partly depend on the understanding of the term *model*. Using some model-theoretic results, he shows a certain weakness of the notion of *categoricity in power*. Then he discusses a notion of categoricity based on models with an absolute interpretation of set-theoretical concepts. Grzegorczyk introduces a new, very important notion of a *constructive model*, using the well-known procedure of Skolemization. He calls a theory *constructively categorical*, if any two of its constructive models are isomorphic. Then he proves the theorem about the existence of constructive models (based on terms). He also gives a necessary and sufficient condition for a theory to be constructively categorical. Grzegorczyk proves that:

1. Robinson's arithmetic has a constructively categorical axiom system (the standard natural numbers form its constructive model).

2. The complete system of the arithmetic of real numbers has a constructively categorical axiom system (its constructive model is isomorphic to the set of all real algebraic numbers). Also the complete system of Euclidean geometry is constructively categorical.

Grzegorczyk introduces two kinds of non-categorical theories. A theory T is *essentially non-categorical* if and only if it has no consistent extension with a constructively categorical axiom system. A theory T is *weakly essentially non-categorical* if and only if it has no recursively enumerable consistent extension with a constructively categorical axiom system. He proves that the theory of a dense ordering with at least two elements is essentially non-categorical. He also proves some more general theorems concerning the notions in question.

Finally, the author considers yet another notion of categoricity: a theory T is called μ-categorical if and only if it has a minimal model and any two of its minimal models are isomorphic. Here by a *minimal model* of a theory T one means any model $\mathfrak{M}$ of T such that if $\mathfrak{N}$ is a model of T and a submodel of $\mathfrak{M}$, then $\mathfrak{M}$ and $\mathfrak{N}$ are isomorphic. Grzegorczyk proves that:

1. The complete theory of dense ordering is μ-categorical.
2. Every minimal model is at most denumerable.
3. Any theory which has no finite models and has minimal models and is categorical in power $\aleph_0$ is μ-categorical.
4. The elementary arithmetic of natural numbers is μ-categorical.

In model theory one considers several special models: atomic, prime, saturated, homogeneous, universal, and so on. Intuitively speaking, atomic models are *poor* and saturated models are *rich* – both these conditions concern the number of types realized in these models. It seems that both of these notions are unrelated to the concepts of a standard model and intended model. However, they are very useful in model theory. For instance, they can be used in the proof of the theorem (Vaught) stating that a complete theory cannot have exactly two non-isomorphic models.

It is not surprising that one can obtain characterization results in strong systems of logic. As an example let us mention Scott's isomorphism theorem characterizing countable models of the infinitary language $L_{\omega_1\omega}$:

> For any countable model $\mathfrak{A}$ of this language there exists a sentence φ of $L_{\omega_1\omega}$ such that for all countable models $\mathfrak{B}$ of this language: $\mathfrak{B} \models \varphi$ if and only if $\mathfrak{B}$ is isomorphic with $\mathfrak{A}$.

4.7 Cognitive aspects

When one is willing to talk about intended models, then it is understood that one has well-developed *mathematical intuitions* about it. The very term *mathematical intuition* may be misleading; it embraces many (only partly comparable) topics. We think that at least three stages in the development of such intuitions should be distinguished:

1. *Proto-intuitions.* Connected with our cognitive abilities: e.g. subitization or distinguishing inner and outer parts of an object, approximate evaluation of size, recognition of rhythm, and so on.
2. *Intuitions learned at school.* These intuitions are imposed on us by the symbolic violence in the process of education. They include: cardinal

and ordinal aspects of numbers, abilities to perform prescribed algorithms, fluency in symbol manipulation, and so on.

3. *Intuitions of professional mathematicians.* They are hard to describe and may very well be idiosyncratic in character, depending on individual talent and abilities. They can be identified by a study of the source texts.

Intuitions about intended models are collected at all these stages. It is tacitly assumed in the theoretical investigations in the foundations of mathematics that intended models should be connected with the last of the stages mentioned above.

We have proposed the addition of one new context to the old dichotomy *context of discovery – context of justification*:

1. Context of discovery: mathematical intuition. Mathematical results are obtained as a result of intuitive thinking which only later becomes justified by the corresponding proofs.

2. Context of justification: proof (in the mathematical sense). This covers the justifications of mathematical results as they are presented in publications. Such justifications have considerable stability but one should also remember that opinions as to what constitutes legitimate methods of proof are changing (just think about the axiom of choice or computer-assisted proofs).

3. Context of transmission: intuitive explanations. This context embraces the ways in which mathematical knowledge and abilities based on it are transmitted. Hence, it covers the learning and teaching of mathematics as well as the ways in which mathematics is being presented to the general public. A special role here is played by *intuitive explanations* which transgress the formal deductive exposition and may be thought of as auxiliary tools helpful as far as a better understanding of mathematical ideas is concerned.

4.8 Final remarks

The following factors seem to be important in the context of intended and standard models:

1. *Collected data (including proven theorems).* The more we know about investigated structures (for instance numbers or space representations)

the better we understand them. The objects of mathematics become domesticated not only by the theorems we can prove about them but also by their application (together with these theorems) in science.

2. *Mathematical intuitions.* These factors refer to our cognitive abilities as well as to tradition, that is the way mathematics has been cultivated through the millennia.

3. *Mathematical arguments.* These arguments concern the categoricity (or categoricity in power) and completeness of theories. They depend on the background logical system. Isomorphism theorems in algebra mentioned above belong here as well as Zermelo's isomorphism theorems for his system of set theory from 1930. These arguments are most important for mathematicians themselves.

4. *Points of view.* Harvey Friedman wrote about the consequences of accepting different *points of view* in set theory, such as the Borel, predicative, constructible point of view. According to his claim, incompleteness results are consequences of the acceptance of very general abstract sets. If we limit ourselves for instance to Borel sets only, then continuum hypothesis can be proven for them. The above points of view are connected with mathematical research practice: it is unusual to deal with very sophisticated pathological objects in everyday work.

5. *Philosophical justification.* Such arguments may refer to pragmatically based decisions (as in the case of maximal axioms in set theory) or even to the aesthetic values thought of by many mathematicians as the guiding principles of their research work. It is highly likely that much more can be said about the philosophical assumptions underlying mathematical research, but we shall stop here, and set this topic aside for a future study.

Intended models of mathematical theories are determined as a result of mathematical research practice. Standard models are determined only on the metatheoretical level. The decision to call a model standard involves an irremovable pragmatic component.

Bibliography

Awodey, S. Reck, E.H. 2002a. Completeness and categoricity. Part I: Nineteenth-century axiomatics to twentieth-century metalogic. *History and Philosophy of Logic* 23, 1–30.

Awodey, S., Reck, E.H. 2002b. Completeness and categoricity. Part II: Twentieth-century metalogic to twenty-first-century semantics. *History and Philosophy of Logic* 23, 77–94.

Baldwin, J. 2015. Axiomatizing changing conceptions of the geometric continuum I: Euclid-Hilbert. Available at:

http://homepages.math.uic.edu/~jbaldwin/pub/axconIsub.pdf

Baldwin, J. 2015. Axiomatizing changing conceptions of the geometric continuum II: Archimedes-Descartes-Hilbert-Tarski. Available at:

http://homepages.math.uic.edu/~jbaldwin/pub/axconcIIfeb29.pdf

Bell, J.L. 2005. *The continuous and the infinitesimal in mathematics and philosophy*. Available at:

http://publish.uwo.ca/~jbell/TheContinuous.pdf

Boyer, C.B. 1949. *The history of the calculus and its conceptual development*. Dover Publications, Inc., New York.

Brouwer, L.E.J. 1927. Über Definitionsbereiche von Funktionen. *Mathematische Annalen* 97, 60–75.

Button, T., Smith, P. 2012. The philosophical significance of Tennenbaum's theorem. *Philosophia Mathematica* 20, 114–121.

Button, T., Walsh, S. 2016. Structure and categoricity: determinacy of reference and truth-value in the philosophy of mathematics. ArXiv: 1501.00472v2

Carnap, R. Bachmann, F. 1936. Über Extremalaxiome. *Erkenntnis* 6, 166–188. Translated by H.G. Bohnert: Carnap, R. Bachmann, F. 1981. On extremal axioms. *History and Philosophy of Logic* 2, 67–85.

Corcoran, J. 1980. Categoricity. *History and Philosophy of Logic* 1, 187–207.

Corcoran, J. 1981. From categoricity to completeness. *History and Philosophy of Logic* 2, 113–119.

Dedekind, R. 1872. *Stetigkeit und Irrationale Zahlen*. Vieweg, Braunschweig.

Di Giorgio, N. 2010. *Non-standard models of arithmetic: a philosophical and historical perspective*. MSc Thesis, Universiteit van Amsterdam, Institute for Logic, Language and Computation.

Ehrlich, P. 1995. Hahn's *Über die nichtarchimedischen Grössensysteme* and the development of the modern theory of magnitudes and numbers to measure them. Available at:

http://www.ohio.edu/people/ehrlich/HahnNew.pdf

Ehrlich, P. 2006. The rise of non-Archimedean mathematics and the roots of a misconception I: The emergence of non-Archimedean systems of magnitudes. *Archive for the History of Exact Sciences* 60, 1–121.

Ehrlich, P. 2012. The absolute arithmetic continuum and the unification of all numbers great and small. *The Bulletin of Symbolic Logic* 18 (1), 1–45.

Feferman, S. 2008. Conceptions of the continuum. Available at:

https://math.stanford.edu/~feferman/papers/ConceptContin.pdf

Ferreirós, J. 1999. *Labyrinth of thought. A history of set theory and its role in modern mathematics.* Birkhäuser Verlag, Basel, Boston, Berlin.

Fraenkel, A.A. 1928. *Einleitung in die Mengenlehre.* Verlag von Julius Springer, Berlin.

Fraenkel, A.A., Bar-Hillel, Y., Levy, A. 1973. *Foundations of set theory.* North-Holland Publishing Company, Amsterdam London.

Friedman, H. 1992. The incompleteness phenomena. *Proceedings of the AMS Centennial Symposium, August 8–12, 1988.* American Mathematical Society, 49–84.

Gaifman, H. 2004. Nonstandard models in a broader perspective. In: Enayat, A., Kossak, R. (Eds.) *Nonstandard models in arithmetic and set theory.* AMS Special Session Nonstandard Models of Arithmetic and Set Theory, January 15–16, 2003, Baltimore, Maryland, *Contemporary Mathematics* **361**, American Mathematical Society, Providence, Rhode Island, 1–22.

Gödel, K. 1940. The consistency of the axiom of choice and of the generalized continuum hypothesis with the axioms of set theory. *Annals of Mathematics Studies* 3, Princeton.

Grassmann, H. 1861. *Lehrbuch der Arithmetik.* Verlag von Th. Chr. Fr. Enslin, Berlin.

Grzegorczyk, A. 1962. On the concept of categoricity. *Studia Logica*, 13: 39–66.

Hahn, H. 1907. Über die nichtarchimedischen Grössensysteme. *Sitzungsberichte der Kaiserlichen Akademie der Wissenschaften, Wien, Mathematisch-Naturwissenschaftliche Klasse* **116** (Abteilung IIa), 601–655.

Halbach, V., Horsten, L. 2005. Computational structuralism. *Philosophia mathematica* 13 (2), 174–186.

Hilbert, D. 1988. *Foundations of Geometry*. Open Court, Lasalle.

Hintikka, J. 1986. Extremality assumptions in the Foundations of Mathematics. *Philosophy of Science Association* 2, 247–252.

Hintikka, J. 1991. Carnap, the universality of language and extremality axioms. *Erkenntnis* 35 (1–3), 325–336.

Lawvere, F.W. 1979. Categorical dynamics. In: *Topos Theoretic Methods in Geometry*, Aarhus Math. Inst. Var. Publ. series 30.

Lindenbaum, A., Tarski, A. 1936. Über die Beschränkheit der Ausdruckmittel deduktiver Theorien. *Ergebnisse eines mathematischen Kolloquiums* 1934–1935, 7, 15–22.

Mostowski, A. 1967. Recent results in set theory. In: Lakatos, I. (Ed.) *Problems in the Philosophy of Mathematics*. North Holland Publishing Company, Amsterdam, 82–96, 105–108.

Myhill, J. 1952. The hypothesis that all classes are nameable. *Proc. Nat. Acad. Sci. USA* 38, 979.

Neumann, J. von 1925. Eine Axiomatisierung der Mengenlehre. *Journal für die reine und angewandte Mathematik* 154, 219–240.

Pogonowski, J. 2010. Remarks on intended models of mathematical theories. *Lingua Posnaniensis* 53(2), 83–100.

Pogonowski, J. 2019. *Extremal axioms. Logical, mathematical and cognitive aspects*. Wydawnictwo Nauk Społecznych i Humanistycznych UAM, Poznań.

Putnam, H. 1980. Models and reality. *The Journal of Symbolic Logic* 45, 464–482.

Quinon, P., Zdanowski, K. 2007. The intended model of arithmetic. An argument from Tennenbaum's theorem. Available at:

http://www.paulaquinon.com/wp-content/uploads/2013/04/QuinonZdanowski.pdf

Schiemer, G. 2010a. Fraenkel's axiom of restriction: axiom choice, intended models, and categoricity. In: Löwe B., T. Müller, T., (Eds.) *Philosophy of mathematics: sociological aspects and mathematical practice*. College Publications, London, 307–340.

Schiemer, G. 2010b. *Carnap's early semantics*. PhD Dissertation, Universität Wien.

Schiemer, G. 2012. Carnap on extremal axioms, "completeness of models", and categoricity. *The Review of Symbolic Logic* 5 (4), 613–641.

Schiemer, G. 2013. Carnap's early semantics. *Erkenntniss* 78 (3), 487–522.

Schiemer, G., Reck, E.H. 2013. Logic in the 1930s: type theory and model theory. *The Bulletin of Symbolic Logic* 19 (4), 433–472.

Schiemer, G., Zach, R., Reck, E.H. 2015. Carnap's early metatheory: scope and limits. Available at: arXiv:1508.05867v1.pdf

Skolem, T. 1970. *Selected works in logic*. Edited by Jens Erik Fenstad. Universiteitsforlaget, Oslo - Bergen - Tromsö.

Suszko, R. 1951. Canonic axiomatic systems. *Studia Philosophica* 4, 301–330.

Tarski, A. 1940. On the completeness and categoricity of deductive theories. Appendix in Mancosu, P. 2010. *The adventure of reason. Interplay between philosophy and mathematical logic, 1900–1940*, Oxford University Press, Oxford, 485–492.

Weyl, H. 1918. *Das Kontinuum. Kritische Untersuchungen über die Grundlagen der Analysis*. Veit, Leipzig.

Zermelo, E. 1930. Über Grenzzahlen und Mengenbereiche: Neue Untersuchungen über die Grundlagen der Mengenlehre. *Fundamenta Mathematicae* 16, 29–47.

Development in Humanities

edited by Zbigniew Drozdowicz (General Editor) and Sławomir Sztajer
(Adam Mickiewicz University, Poznán/Poland)

Zbigniew Drozdowicz
Higher Education Institutions in the Process of Transformation and Modernization
Global and Local Issues
This book is dedicated to the problems faced by universities. The author frequently refers to those events from the past that resulted in universities becoming institutions of public benefit. This benefit is of course understood in various ways, but in ways always involving the institutions' function of serving. What is debatable is whom and what they were and are meant to serve, and how they can and how they should fulfil these functions. Although such questions are global in character, the answers to them can be both global and local, meaning that they may relate to both the most general tasks of universities and to those that might be or are to some degree only performed by institutions of a particular type.
Bd. 16, 2018, 210 S., 29,90 €, br., ISBN 978-3-643-91058-5

Slawomir Sztajer; Jarema Drozdowicz
Changing Trajectories of Religion and Popular Culture
Cognitive and Anthropological Dimensions
Cultural diversity and cultural change make it difficult to define and theorize cultural phenomena. This is especially apparent in the case of such cultural areas as religion and popular culture. This book presents ways to understand and explain the diversity and variability of religious and popular culture phenomena. The first part of this book focuses on the cognitive foundations and cultural dimensions of religious phenomena. The cognitive science of religion provides a new theoretical framework for explaining religious diversity and variability. The second part is dedicated to the study of selected phenomena of popular culture from the perspective distinctive to cultural anthropology. It attempts to bring into light this features of popular culture phenomena that have direct impact on cultural subjects.
Bd. 15, 2018, 162 S., 29,90 €, br., ISBN 978-3-643-91056-1

Ferdinand Fellmann
The Couple
Intimate Relations in a New Key
This book is about the mating couple – considered as both the origin and the foundation of a strong social self. Despite increasing self-realisation, the individual always is in need of being justified through an intimate relationship in which both partners accept one another such as they are. In view of the process of global urbanization empathy and reliability are the royal road to overcoming existential loneliness.
Bd. 14, 2016, 164 S., 29,90 €, br., ISBN 978-3-643-90770-7

Zbigniew Drozdowicz
Essays on European Liberalism
History and Ideology
The historical controversies and misunderstandings arising between the liberals and their numerous and socially important adversaries persuaded the author of the book to survey the traditions and contemporary ideological frame of liberalism. One of the key motives behind this endeavour has been the attempt to demonstrate that there is no single liberalism, and in fact never has been. There have been, however, multiple different liberalisms, and so it remains important to contextualise them in the social and cultural contexts of particular continents as well as specific countries.
Bd. 13, 2013, 176 S., 29,90 €, br., ISBN 978-3-643-90433-1

Andrzej Wiśniewski
Essays in Logical Philosophy
Bd. 12, 2013, 120 S., 29,90 €, br., ISBN 978-3-643-90381-5

Józef Baniak
The Image of the Priest in the Awareness of Polish Youth
A Sociological Study
Bd. 11, 2013, 168 S., 29,90 €, br., ISBN 978-3-643-90380-8

Pawel Zeidler
Models and Metaphors as Research Tools in Science
A Philosophical, Methodological and Semiotic Study of Science
Bd. 10, 2013, 192 S., 29,90 €, br., ISBN 978-3-643-90379-2

Andrzej Przyłebski
Sense, Meaning and Understanding
Towards a Systematic Hermeneutical Philosophy
Bd. 9, 2013, 176 S., 29,90 €, br., ISBN 978-3-643-90378-5

Marian Golka
The Sociology of the Artist in the Postmodern Era
Pride and Uncertainty
Bd. 8, 2013, 176 S., 29,90 €, br., ISBN 978-3-643-90377-8

Zbigniew Drozdowicz
Rationality Standards of Social Life
Traditions and the Present
Bd. 7, 2013, 192 S., 29,90 €, br., ISBN 978-3-643-90376-1

Ewa Nowak
Experimental Ethics
A Multidisciplinary Approach
Bd. 6, 2013, 168 S., 29,90 €, br., ISBN 978-3-643-90375-4

Ewa Rewers
The Contradictions of Urban Art
Contrasting Models of Critical Consciousness
Bd. 5, 2013, 184 S., 29,90 €, br., ISBN 978-3-643-90374-7

Zbigniew Woźniak
The Elderly in Social Policy
Towards a New Architecture of Gerontological Programmes
Bd. 4, 2013, 184 S., 29,90 €, br., ISBN 978-3-643-90373-0

Grzegorz Dziamski
Art in Postmodern Era
From a Central European Perspective
Bd. 3, 2013, 136 S., 29,90 €, br., ISBN 978-3-643-90372-3

Zbigniew Drozdowicz
Standards of Religious Rationality
Tradition and Modern Times
Bd. 2, 2013, 160 S., 29,90 €, br., ISBN 978-3-643-90389-1

Zbigniew Drozdowicz
Standards of Philosophical Rationality
Traditions and Modern Times
Bd. 1, 2013, 144 S., 29,90 €, br., ISBN 978-3-643-90388-4

LIT Verlag Berlin – Münster – Wien – Zürich – London
Auslieferung Deutschland / Österreich / Schweiz: siehe Impressumsseite